PRIMER ON ENGINEERING STANDARDS

EXPANDED TEXTBOOK EDITION

Wiley-ASME Press Series List

PRIMER ON ENGINEERING STANDARDS

EXPANDED TEXTBOOK EDITION

Owen R. Greulich

Vienna, Virginia, USA

Maan H. Jawad

Global Engineering & Technology
Camas, Washington, USA

This Work is a co-publication between ASME Press and John Wiley & Sons, Ltd.

© 2018 ASME

This Work is a co-publication between ASME Press and John Wiley & Sons Ltd

The right of Owen R. Greulich and Maan H. Jawad to be identified as the authors of this work has been asserted in accordance with law.

Registered Offices
John Wiley & Sons, Inc., 111 River Street, Hoboken, NJ 07030, USA
John Wiley & Sons Ltd, The Atrium, Southern Gate, Chichester, West Sussex, PO19 8SQ, UK

Editorial Office
The Atrium, Southern Gate, Chichester, West Sussex, PO19 8SQ, UK

For details of our global editorial offices, customer services, and more information about Wiley products visit us at www.wiley.com.

Wiley also publishes its books in a variety of electronic formats and by print-on-demand. Some content that appears in standard print versions of this book may not be available in other formats.

Library of Congress Cataloging-in-Publication Data applied for

ISBN: 9781119466178

Cover Design: Wiley
Cover Image: © Anthony Kyriazis/Gettyimages

Set in 10/12pt and TimesLTStd by Spi Global, Chennai, India

Printed and bound by CPI Group (UK) Ltd, Croydon, CR0 4YY

10 9 8 7 6 5 4 3 2 1

*To those who seek excellence through
their knowledge of standards*

Contents

Preface

Engineering principles including classical and numerical analysis as well as other engineering techniques are essential for the engineer to perform various designs. However, as society gets more interdependent and the common implements of daily life more complex and sophisticated, standards become more and more indispensable as additional engineering tools. This book introduces the concept of standards as well as their impact and value. It includes a brief history of standards and it addresses the different ways in which they come about. Some of the chapters discuss the role of government in creating standards as well as the processes by which nongovernmental standards are produced. Other chapters discuss the different types and applications of standards, how interpretations of standards are obtained, the problem of how to ensure conformity with standards, and what might be done when conformity cannot be attained.

Some characteristics of a "good standard" are presented, along with some pitfalls to avoid in producing a standard. Benefits of getting involved in the standards development process are explained, along with pointers on both selecting a standards organization to get involved with and how to go about it. This book provides a short synopsis of "Standards" to enable the reader get a quick understanding of the various aspects, ramifications, and implications of standards. It consists of eleven chapters and four appendices. Various case studies are included to help the reader develop an in-depth understanding of the topics discussed. The wide range of topics covered in this book is intended to give the reader a good starting point in understanding how standards play an integral part of the engineering profession.

There are tens, perhaps hundreds of thousands of engineering standards worldwide, covering every imaginable subject related to engineering. Listing them all would be a monumental undertaking and this book, by necessity, covers only a small portion of them. The appendices at the end of this book provide assistance in identifying a few of these engineering standards, who developed and maintains them, and contact information to help the reader obtain further information.

April, 2018   Owen R. Greulich
                                                                  Washington, DC

                                                                  Maan H. Jawad
                                                              Camas, Washington

Acknowledgments

The authors acknowledge Mark Jawad for providing background information on the USB standard and Kevin Jawad for providing history regarding computer development.

We also acknowledge Cathy Greulich and Dixie Jawad for their patience while the authors were writing the manuscript, and special thanks are extended to Mary Grace Stefanchik and to all the other ASME and Wiley personnel for their valuable assistance and guidance on the book.

1

Introduction

(Courtesy of XKCD, www.xkcd.com)

1.1 Background

Standards, including procedures, rules, codes, regulations, and jurisdictional requirements, play an important role in the engineering world. They furnish a means of ensuring consistent designs, quality, and operating characteristics, with adequate reliability, safe operation of components, and well-defined configurations. This book details their development, applications, limitations, and benefits to give the user, specifier, and standard writer a proper perspective of their importance, usefulness, and limitations, to promote their effective use and to improve the quality of future standards. While focusing on standards

Primer on Engineering Standards: Expanded Textbook Edition, First Edition.
Owen R. Greulich and Maan H. Jawad.
© 2018, The American Society of Mechanical Engineers (ASME), 2 Park Avenue, New York, NY, 10016, USA (www.asme.org). Published 2018 by John Wiley & Sons Ltd.

themselves, it also provides a brief introduction to the field of conformity assessment, by which compliance with standards is assured and verified.

This book does not attempt to address legal, commercial, or other nonengineering aspects of standards.

Rules, procedures, and standards can be developed by a single individual within or with authority over an organization or operation, by a subgroup of an organization, by the organization as a whole, or by other groups with a common interest.

The reader will find that common usage of terminology related to standards is inconsistent in the literature. Accordingly, the following definitions are provided and will be used in this book:

Rule: A single specific requirement that must be met. Many types of such requirements exist, such as requirements to perform actions, for how to perform actions, for results that must be achieved, for specific properties or characteristics that must be attained, and for dimensions that must be met.

Procedure: A set of rules regarding how a task or function is performed. Procedures are used to ensure consistency of results and to promote efficiency.

Standard: A set of rules and/or procedures recognized as authoritative in a particular area of interest.

One definition of standards given by the American Society of Mechanical Engineers (ASME) is as follows:

> *A set of technical definitions, instructions, rules, guidelines, or characteristics set forth to provide consistent and comparable results, including:*
>
> - *Items manufactured uniformly, providing for interchangeability.*
> - *Tests and analyses conducted reliably, minimizing the uncertainty of the results.*
> - *Facilities designed and constructed for safe operation.*

It is of interest to note that by custom, some standards are called codes.

The effectiveness of standards in conducting business is best explained by the American Society for Testing and Materials (ASTM) in a 1991 report:

> *Standards are the vehicle of communication for producers and users. They serve as a common language, defining quality and establishing safety criteria. Costs are lower if procedures are standardized; training is also simplified.*

While the importance of standards is well recognized and while there is what might be called the "legal" definition, as provided in the National

Technology Transfer and Advancement Act (NTTAA) (see below), even the National Institute of Standards and Technology, a part of the US Department of Commerce and that part of the US Government most directly charged with standards development, coordination, and quality, approaches the definition of standards by offering multiple explanations from different sources.

The following text provides the definition from the NTTAA of a standard (clearly covering more than just the engineering standards that are the subject of this book):

> *DEFINITION OF TECHNICAL STANDARDS – As used in this sub-section, the term 'technical standards' means performance-based or design-specific technical specifications and related management systems practices.*

The Office of Management and Budget OMB Circular A-119 further amplifies upon this:

(a) The term "standard," or "technical standard," (hereinafter "standard") as cited in the NTTAA, includes all of the following:

    i. common and repeated use of rules, conditions, guidelines or characteristics for products or related processes and production methods, and related management systems practices;

    ii. the definition of terms; classification of components; delineation of procedures; specification of dimensions, materials, performance, designs, or operations; measurement of quality and quantity in describing materials, processes, products, systems, services, or practices; test methods and sampling procedures; formats for information and communication exchange; or descriptions of fit and measurements of size or strength; and

    iii. terminology, symbols, packaging, marking or labeling requirements as they apply to a product, process, or production method.

(b) The term "standard" does not include the following:

    i. professional standards of personal conduct; or

    ii. institutional codes of ethics.

(c) "Government-unique standard" is a standard developed by and for use by the Federal government in its regulations, procurements, or other program areas specifically for government use (i.e., it is not generally used by the private sector unless required by regulation, procurement, or program participation). The standard was not developed as a voluntary consensus standard as described in Sections d and e.

(d) *"Voluntary consensus standard" is a type of standard developed or adopted by voluntary consensus standards bodies, through the use of a voluntary consensus standards development process as described in Chapter 3. These bodies often have intellectual property rights (IPR) policies that include provisions requiring that owners of relevant patented technology incorporated into a standard make that intellectual property available to implementers of the standard on nondiscriminatory and royalty-free or reasonable royalty terms (and to bind subsequent owners of standards essential patents to the same terms). In order to qualify as a "voluntary consensus standard" for the purposes of this Circular, a standard that includes patented technology needs to be governed by such policies, which should be easily accessible, set out clear rules governing the disclosure and licensing of the relevant intellectual property, and take into account the interests of all stakeholders, including the IPR holders and those seeking to implement the standard.*

(e) *"Voluntary consensus standards body" is a type of association, organization, or technical society that plans, develops, establishes, or coordinates voluntary consensus standards using a voluntary consensus standards development process that includes the following attributes or elements:*

    i. Openness: *The procedures or processes used are open to interested parties. Such parties are provided meaningful opportunities to participate in standards development on a non-discriminatory basis. The procedures or processes for participating in standards development and for developing the standard are transparent.*

    ii. Balance: *The standards development process should be balanced. Specifically, there should be meaningful involvement from a broad range of parties, with no single interest dominating the decision-making.*

    iii. Due process: *Due process shall include documented and publically available policies and procedures, adequate notice of meetings and standards development, sufficient time to review drafts and prepare views and objections, access to views and objections of other participants, and a fair and impartial process for resolving conflicting views.*

    iv. Appeals process: *An appeals process shall be available for the impartial handling of procedural appeals.*

    v. Consensus: *Consensus is defined as general agreement, but not necessarily unanimity. During the development*

> *of consensus, comments and objections are considered*
> *using fair, impartial, open, and transparent processes.*

Some definitions of standards include only those standards with which compliance is voluntary, referring to mandatory standards as technical regulations or by other names. Because many voluntary consensus standards have been incorporated into laws and regulations, confusing the meaning of voluntary, this book does not make this distinction.

Other definitions require consensus, or establishment or approval by a recognized body. These distinctions are often significant, but would exclude such important documents as the standards of the American Boiler Manufacturers Association and the Tubular Exchanger Manufacturers Association (TEMA) as well as certain standards developed by private corporations but well recognized as authoritative (An example of the latter is a series of material and other standards developed by the AO Smith Company and used by them and other companies in the construction of many thousands of pressure vessels beginning in the 1940s and continuing into the 1960s.).

Conformity Assessment: Processes used to verify the compliance of a product, service, person, process or system to either a standard or a regulation (e.g., testing, certification, inspection).

1.2   Procedures and Rules

Procedures and rules are usually developed within an organization to establish operating methods that will lead to consistent desired results. They include such items as drawing and calculation formats, dimensional standards, checking sequences, and hierarchical progression of a task within the organization. The applications of procedures and rules form the operating norm of an organization, and they differ from one organization to another. Hence, the procedures and rules used for the design and manufacturing of the same product at two companies may differ substantially even though the end product is the same.

One company manufacturing a light switch, for example, might bring in large quantities of component parts from suppliers and depend on a high level of automation and process control to ensure a consistent product. A competitor may choose to manufacture all parts in-house, use hand assembly, and control product quality through a rigorous inspection process. The final products may be practically indistinguishable in spite of procedures that have little in common.

Procedures and rules, by their general nature, are limited in scope to an individual task or component within an organization such as a how to manufacture and assemble a gear box, or a methodology for project progression within the organization. A separate procedure then details the next step, whether packaging the gearbox for shipping or assembling it into an automobile. Procedures and rules also can be updated or revised frequently to fit the changing requirements of the organization.

1.3 Standards

1.3.1 History and Purpose of Standards

Any discussion of the purpose of standards is nearly impossible without a concurrent look at their history. The nature of society, the natural human resistance to accept constraints, the thought and effort needed to develop standards, and general inertia have dictated that they not be produced and imposed without a reason. As society has become more complex, the need for standards has occurred more frequently, and the increasing sophistication of society has been mirrored by an increasing sophistication in the standards developed.

While there is sometimes resistance to their development and implementation, the term "Voluntary Consensus Standard" refers to the benefits that standards provide to business and to society in general, and it is applicable to a vast number of standards developed for a wide range of reasons and applications. People, companies, and other organizations now generally recognize that standards can have a favorable effect on their lives, the quality of their work, and their business opportunities. Companies and individuals are therefore willing to apply resources, to give up a certain amount of freedom, to admit that they may function better and be more successful operating as a part of a larger whole than with complete independence, and to share a certain amount of what may be proprietary knowledge. These things are given up in the interest of a safer home or society, efficiency in design, improved business opportunities, a certain amount legal protection in case of product failure, and other benefits.

While standards can be categorized in a number of ways, for purposes of this chapter we will consider the following categorization on the basis of benefits: safety and reliability, quality, uniformity, cost reduction, increased flexibility, variety control, promotion of business, and generally helping society to function. Most standards provide more than one of these benefits.

1.3.2 A Few Examples of Standards throughout the Ages

One of the first known, and very rudimentary, standards, known as the Code of Hammurabi, was developed approximately 4000 years ago, apparently to ensure fairness in the kingdom of Babylonia. In this case, the standard was promulgated by the king and enshrined in laws. Many of the laws included in this document related to crimes, torts, marriage, and general legal obligations. The Code of Hammurabi may be most well-known for its "an eye for an eye and a tooth for a tooth" approach to justice, but portions of this document also provide very basic performance standards for construction of buildings and boats. That is, the document specified what must be achieved. Walls must not fall down, and boats must be tight.

It appears that there was a problem with the quality of workmanship in the kingdom of Babylonia, but the solution had no specific criteria for how things were to be constructed. Rather, the standard that was put in place simply required that the construction be good, and if it failed, specified the penalty. This approach, using what are referred to as performance standards, typically provides little or no guidance as to how the requirements are to be met, simply specifying the required result. A performance standard is in some cases the easiest to write, and it allows the maximum level of flexibility to the implementer, since any means of accomplishing the end is sufficient. This particular standard promoted a fair and just – if somewhat brutal – society. It should also be noted that this standard dealt with very limited aspects of the products, and including in a performance standard all the details that are needed to ensure a successful product can be challenging.

Another step on the way to modern standards came about as a result of a massive fire that burned most of the city of London in 1666. The Rebuilding London Acts, promulgated by the Parliament of England over the following several years, are precursors of current building and safety codes. The specific motivation of these acts was first to ensure safety and stability in society, and they did so by widening streets and by specifying brick construction, so as to prevent recurrence of the disastrous fire.

Compared to the Code of Hammurabi, these Acts are quite prescriptive, specifying street widths, brick construction, thickness of walls (in terms of bricks), story heights and maximum heights of houses, and requirements for roof drainage. The effect of these and other associated requirements was to improve access, provide fire breaks in case another fire got started, and reduce the probability of its spread by replacing what was previously almost entirely wooden construction largely with bricks. Safety was enhanced, and it is to be expected that quality of life may also have been improved.

With the advent of the industrial revolution, other benefits of standards became obvious. As society developed the ability to produce products in quantity and, as machine tools were developed, the interchangeability of components became desirable. In the mid-1800s, for example, the British Standard Whitworth thread system was developed to allow for interchangeability of threaded parts.

Throughout the 1800s, there were many boiler explosions [6]. While mourned, these seemed to be somewhat accepted as a cost of having boilers, which were providing benefits to society in the forms of more efficient transportation, working efficiency, and heating. The explosion of the boiler in a shoe factory in Brockton, MA, in 1905, which resulted in close to 60 deaths, and another shoe factory boiler explosion in Lynn, MA, the following year led to a greater concern with industrial safety. In 1907, the Massachusetts legislature passed the Massachusetts Boiler Law. This was followed by the first ASME Boiler Code in 1914.

The years between then and now have seen a proliferation of standards. These include the further development of the ASME Boiler Code (later to become the Boiler and Pressure Vessel Code (BPVC)) with sections on various types of pressure vessels, materials, welding, inspection, etc., and piping codes, lifting devices standards, electrical codes, and more. More recent work includes standards in the fields of energy efficiency, electronic components, software development, assessment of risk, and conformity assessment, and every time a new technology arises, it seems that standards for its application are not far off.

1.3.3 Classification

Engineering standards are sets of rules and procedures developed, documented, approved by general consensus, and configuration managed to assure the adequacy of a given product, methodology, or operation. Standards are normally developed in committees by people who have experience in a particular field of endeavor and who have an interest in the outcome that the standard is intended to ensure.

There are many ways of looking at and classifying standards. These include

1. performance versus design (descriptive, or prescriptive)
2. mandatory versus voluntary
3. purpose
4. intended user group
5. the way they were developed.

The International Organization for Standardization (ISO) and the International Electrotechnical Commission (IEC) Guide 2, Standardization and related activities – General vocabulary, offers a number of ways of looking at standards in addition to providing vocabulary such as the following:

1. *Level of standardization, which is a generally geopolitical classification based on the coverage of the standard.*
2. *Aims of standardization, addressing issues from fitness for purpose and safety, through protection of the environment.*
3. *Category of the standard, such as technical specifications (e.g., ASME B16.5), codes of practice (e.g., ASME Boiler and Pressure Vessel Code), and regulations (e.g., OSHA and DOT regulations).*
4. *Type of standard, from terminology to testing, product, process, service, interface, and data to be provided.*

With many possible approaches, we will first look at standards as they fit into the following three categories:

1. Limited consensus
2. General consensus (also referred to as Voluntary Consensus Standards)
3. Governmental action.

1.3.4   Limited Consensus Standards

Limited consensus standards, by their nature, are developed by experts in a given organization and then made available to the organization for its guidance. Prime examples are some internal company standards for work hardened, heavy wall, stainless steel tubing for ultrahigh pressure applications, and a number of company standards for compression fittings. Trade groups such as the Heat Exchange Institute (HEI) and the TEMA have developed standards specific to their lines of work. While keeping a greater level of control than would be allowed them if their standards were subject to the American National Standards Institute (ANSI) process and requirements, these organizations reap for the company or trade group as well as for society many of the same benefits that they would achieve by following that more public process.

1.3.5   Voluntary Consensus Standards (VCS)

There is a great difference between the procedures involved in writing limited consensus and voluntary consensus standards. Voluntary consensus standards are developed by experts to serve the need of a given industry with substantial input from other interested parties, and potentially from the public. An example of such standard is the ASME's BPVC. The BPVC contains numerous standards (referred to in the BPVC as sections, divisions, and parts) as will be explained later. The development of each of these standards follows a unique process by which experts from differing interest groups (designers, fabricators, users, and insurers of pressure vessels, and perhaps others) collaborate in writing the rules in order to arrive through a balanced consensus process at a product that is not dominated by any one particular interest group. The resulting standard becomes acceptable to a wide range of interested parties within the industry since all entities were involved in developing it. These standards have been so successful and have garnered such a high level of respect that they are often cited in laws and regulations (a process referred to as "Incorporation by Reference" or "IBR") making them a part of governmental standards, discussed below, and addressed in more detail in Chapter 2.

Typically, in the United States, the organizations developing such standards follow a process certified by the ANSI and note that fact by including "ANSI" in the document name, identifying number, or on their cover or title page. International standards typically follow a similar process. Because of the integrity of the process and the quality of the resulting product, the standards from many

US standards developers are commonly accepted, or even demanded, overseas, and many of the international standards are likewise accepted or required in the United States, both by industry and by jurisdictions.

1.3.6 Governmental Standards

Both jurisdictional and nonjurisdictional standards are produced by a wide range of governmental organizations. Jurisdictional standards are those promulgated by governmental agencies to implement laws, ordinances, and other legal documents, and carrying with them legal enforceability by fines, injunctions, imprisonment, or other sanctions. Nonjurisdictional governmental standards are developed by governmental entities for their own use, whether for internal purposes or for contracted or other outside work.

1.3.6.1 Jurisdictional Standards

Jurisdictional standards are normally produced by the enforcing jurisdiction, which has legal authority for their enforcement. Jurisdictions include the US (federal) Government, states, cities, and municipalities, and their defined sub-authorities such as federal agencies, building departments, and districts, with the authority to enforce the law, regulation, or other standard. Standards such as those published by the US Occupational Safety and Hazard Administration (OSHA) are intended to assure safety in the workplace by providing generally prescriptive requirements for the design and operation of various pieces of equipment and products, or for processes performed, within a wide range of occupations. Many standards published by local jurisdictions are intended to do the same.

1.3.6.2 Nonjurisdictional Governmental Standards

These standards are typically produced by governmental entities to promote safe, predictable, and efficient performance of their work. Some of these will be implementing standards for jurisdictional standards. An example of this type of standard would be an internal standard of a large governmental agency providing requirements enforced internally and used to ensure compliance with a jurisdictional standard enforced by another agency such as the Department of Transportation (DOT). For example, it might detail the design process or other specific requirements related to procurement of pressure vessels to be used by the National Science Foundation on research aircraft. Their application could be either internal or through its application on an external procurement.

Another category of nonjurisdictional governmental standards might be used within a National Aeronautics and Space Administration (NASA) Center to ensure reliable and consistent designs for instrumentation on spacecraft. Such a standard could define the design flow, component ranges, and reliability requirements needed for a successful mission. It helps with the implementation of the mission, but is not an implementation of any other governmental mandate such as a regulation from OSHA, DOT, Department of Energy (DOE), or the Environmental Protection Agency (EPA).

1.4 Applicability of Standards

Standards form the backbone for many engineering processes by providing the following requirements and guidelines:

1. Practical limits for operating conditions to improve safety and reliability
2. Permissible materials of construction, performance criteria, and material properties
3. Safe design rules
4. Construction details
5. Available methodologies for inspection and testing
6. Safe operating parameters
7. Process control
8. Practical limits for operating conditions to improve safety and reliability.

Standards often take theoretical material and make modifications to permit its application in a practical way. Theoretical equations for a variety of engineering applications are found in textbooks. An example is the equation for circumferential stress in a thin cylindrical shell ($R/T < 10$), given by

$$S = \frac{PR}{T} \tag{1.1}$$

where S is stress, P is internal pressure, R is inside radius, and T is shell thickness. Another example is the elastic buckling of a column given by the equation

$$S_c = \frac{\pi^2 E}{(L/r)^2} \tag{1.2}$$

where S_c is the critical compressive buckling stress, E is the modulus of elasticity, L is the effective length, and r is the radius of gyration of the column.

While representing a classical analysis and being theoretically correct, both of these equations have limitations that could lead to unsafe designs if not used properly. Accordingly, standards have been written to promote the effective application of engineering principles to everyday life, establishing limits and modifying theoretical equations in order to aid engineers in developing safe designs.

Thus for easy, practical, everyday usage, the ASME Boiler and Pressure Vessel Code [3] modifies Eq. (1.1) to the following:

$$t = \frac{PR}{SE - 0.6P} \tag{1.3}$$

In this equation, t is the required minimum wall thickness, and P is pressure. R is the inside radius and the denominator is an approximation of the Lame's equation for thick shells. S includes a design factor (currently 3.5 in Section VIII, Division 1 of the BPVC) used to account for material properties, including a reasonable design life, to account for unknowns such as a certain amount of out-of-roundness (limited by the Code to 1%), residual stresses, possible local material variability below the specified material properties (from another part of the Code), and other aspects of final construction. Finally, the joint efficiency E provides an additional factor based on the level of nondestructive evaluation performed on welds, and allowing for a certain amount of weld defects.

Similarly, the Steel Construction Manual from the American Institute of Steel Construction (AISC) [1] reformats Eq. (1.2) in order to prevent S_c from getting higher than the yield stress of the material. The reformatting also allows for a certain amount of material variation, lack of straightness of the column and other variation in form, and actual installation characteristics, with, again, a design factor to ensure that the column does not fail just as it reaches its design load. Equation (1.2) is thus modified to the following format

$$F_a = F_n/1.67 \tag{1.4}$$

where

$$F_n = (0.658^{\lambda_c^2})(F_y) \quad \text{when } \lambda_c \leq 1.5 \text{ or } \frac{L}{r} \leq 4.71\sqrt{E/F_y}$$

$$F_n = 0.877\, F_y/\lambda_c^2 \quad \text{when } \lambda_c > 1.5 \text{ or } \frac{L}{r} \leq 4.71\sqrt{E/F_y}$$

$$\lambda_c^2 = F_y/F_e$$

$$F_c = \pi^2 E/(L/r)^2$$

and

$$F_y = \text{yield stress}$$

$$E = \text{elastic modulus}$$

$$F_a = \text{allowable compressive stress}$$

$$L = \text{effective length}$$

$$r = \text{radius of gyration}$$

Often, one of the most important aspects of an engineering standard is the application of a design factor, variously referred to as "design factor," "safety factor," and "ignorance factor."

1.4.1 Permissible Materials of Construction, Performance Criteria, and Material Data

Engineers have access to thousands of materials of construction varying from metallics such as steel, aluminum, copper, and titanium to nonmetallics such as concrete, wood, plastic, and composites. Numerous subcategories of these materials are available with their own properties, limitations, and advantages. Steel, for example, has many product forms such as structural steel shapes, rods, plates, forgings, tubes, and castings, and a myriad of alloys designed to perform under different types of loads and over different temperature ranges.

Engineers rely on national and international material standards to provide them data such as the chemistry, tensile and yield strength, and other physical properties for use in design of components and systems.

ASTM standards provide a wealth of material characteristics at room temperature for thousands of metallic and nonmetallic materials. In addition, ASME provides material data for several thousand metallic materials at elevated temperatures and testing requirements for low-temperature operations.

1.4.2 Safe Design Rules

Engineers must often rely on national and international standards for obtaining appropriate material properties and safety factors for their designs. Such standards establish factors of safety based on experience, practical limits, and understanding of material characteristics such as endurance limits, creep, and fatigue crack growth rates. Most engineers do not have the means of establishing such factors on their own. A case in point is the ASME BPVC, Section VIII, Division 1 allowable stress, referred to in the discussion above. It uses four criteria for establishing the allowable tensile stress. This allowable stress is taken as the smaller of the values obtained from 2/3 times the yield stress, 0.286 times the tensile strength, stress that causes rupture at 100,000 h, or stress at a creep rate of 0.01% in 1000 h. Division 1 also uses a factor of safety of 3 for lateral buckling of cylindrical shells and a factor of safety of 10 for the axial compression of cylindrical shells. Determination of these values is based on a significant amount of underlying work experience, including assumptions used in the derivation of the basic theoretical equations, practical limits, and life calculations. While not obvious in the simple tables providing allowable stress values, significant effort goes into determination of the criteria just noted, including the obvious stress failure criteria, and also fatigue life, comparison to success of other codes, and actual success in service.

1.4.3   Construction Details

Details of construction of intricate components such as column-to-beam connections, heat exchangers, and concrete slab reinforcement are provided in established standards. The details in these standards are based on years of practical experience that are in this way made accessible to the engineer. However, even these standards change sometimes based on new information. A case in point is the considerable changes made in steel construction joint details and other design requirements in construction standards in the years following the Sylmar (1971) and the Northridge (1994) earthquakes in the Los Angeles area.

1.4.4   Available Methodologies for Inspection and Testing

Although some industries depend almost entirely on process control, many components need to be inspected during manufacturing to assure good quality. This is especially true if the manufacturing does not involve mass production. For metallic structures, inspection methods such as radiography, ultrasonic testing, and magnetic particle examination are among those available. These methods are covered by standards produced by ASTM and ASME that provide detailed requirements for the engineer or technician performing the inspection, ensuring consistency in the performance and the results of inspections, and thereby in the final product.

1.4.5   Safe Operating Parameters

Certain precautions must be followed in order to operate equipment and machinery in a safe manner. Standards are available to assist the engineer and operator in their jobs. Some of these are industry standards put in place to protect employees and to reduce employer liability, while others are jurisdictional standards such as the OSHA regulations.

1.4.6   Conformity Assessment

With standards applicable to all fields, and compliance with them critical to guarantee society the benefits that they provide, a whole field has arisen to verify compliance. This field, referred to as *conformity assessment*, is defined as those processes used to verify the compliance of a product, service, person, process or system to either a standard or a regulation (e.g., testing, certification, inspection).

Conformity assessment has its own set of standards, and while manufacturing, service, and other organizations devote significant resources to maintaining compliance, there is also a cadre of organizations (ANSI, ASME, ISO, UL, etc.) that make a business of certifying company processes and compliance. A more detailed discussion of this field is provided in Chapter 8.

1.5 Summary

Standards have existed throughout much of recorded history. While the first ones were fairly rudimentary, with time they have become more sophisticated, more complete, and more pervasive. As society has become more complex and as technology has advanced, they have also become more necessary for the smooth functioning of society. It can only be expected that more standards will continue to be developed, and that they will become more interlinked. A basic understanding of standards will be useful in any technical field.

2

Role of Governments in Standards

Today, more than ever, standards are an imperative undertaking. Standards are the building blocks for innovation and competitiveness. Our nation's ability to compete and lead in a rapidly changing global economy is closely related to our leadership in the development and effective use of standards and standardization processes. Standards provide the common language that keeps domestic and international trade flowing. It is difficult to overestimate their critical value to both the U.S. and global economy.

– Patrick Gallagher, Director,
National Institute of Standards and Technology, 2010

2.1   Overview

There is a long history of governmental involvement with standards, beginning with establishment of some standards by government and continuing through the use by the government of standards developed by others. Facets of this involvement include direct development of standards (for internal or external use), involvement in the development of standards by other organizations, urging by governmental entities for development or change of standards, and use of standards.

For those standards developed by a government, there is also a distinction to be made among those intended for internal use and those for use outside of the government. Among those intended for use outside of the government, there is a further distinction between those with which compliance is mandatory and those produced for the benefit of industry, etc., but for which use is optional.

Primer on Engineering Standards: Expanded Textbook Edition, First Edition.
Owen R. Greulich and Maan H. Jawad.
© 2018, The American Society of Mechanical Engineers (ASME), 2 Park Avenue, New York, NY, 10016, USA (www.asme.org). Published 2018 by John Wiley & Sons Ltd.

Most industrialized nations have formal standards policies that include developing some of their own standards as well as being involved with nongovernmental and intergovernmental standards developing organizations (SDO) such as ASTM International and the International Organization for Standardization (ISO). The United States, Canada, all of the major European nations, most African, South American, and Asian nations, Australia, and members of the Russian Federation are members of ISO.

One of the most important purposes of governmental involvement with standards is to assure safety of products. This involvement can take the form of writing independent standards or requesting additional standards from the private sector. An example is the government's involvement in the aftermath of the United Airlines flight 232 crash in 1989 in Sioux City, Iowa (Fig. 2.1). A catastrophic failure of its tail engine resulted in the loss of the primary as well as the two secondary hydraulic systems, causing flight controls to be inoperative. Although the crash resulted in 111 deaths, the crew's expert handling of the disabled plane based on their United Airlines "Crew Resource Training" minimized substantially the number of fatalities. The government has since mandated similar training for other commercial airlines flying in the United States.

Figure 2.1   Crash of plane, Sioux City, Iowa, 1989
(© Associated Press/John Gaps III)

2.2   History

As noted in section 1.3.2, governments have been involved with standards development for thousands of years. The extent and sophistication of this involvement has increased significantly in parallel with the vast increase in technology and standards development in the private sector.

From the Code of Hammurabi, to the Rebuilding London Acts, to the early boiler laws in the United States, through the evolution to greater use of and involvement in the creation of industry and consensus standards, there has been a continual progression in the number, complexity, and importance of standards.

The same progression that has occurred over hundreds of years in the more developed nations is being compressed into a much shorter period of time in emerging nations. This happens naturally as these economies develop and wish to produce components and products of a quality and consistency that will be accepted by companies and consumers in other countries. The detailed information contained in standards allows, for better or for worse, a much quicker evolution from a preindustrial society to an industrialized, exporting, importing, and consumer society.

2.3   Aspects of Governmental Involvement with Standards

Federal, state, and local governments in the United States, and their counterparts in other countries, are involved with standards in a number of ways. Nearly every governmental entity has standards of some sort that have been developed to facilitate smooth operations, and many have standards intended for use outside of their own organization, both mandatory and nonmandatory. In addition, consistent with stated policies, they are often involved in the development of voluntary consensus standards (VCS) by other organizations (examples in the United States include ASME, ASTM, IEEE, NFPA, etc.). The government also sometimes is involved in standards development by encouraging change in existing VCS or the development of new ones. The United States Chemical Safety Board (CSB), an independent federal organization responsible for investigating the causes of accidents due to hazardous materials in commerce and industry, often suggests increased use of, or changes to, standards in order to increase safety in industries using hazardous materials. Other governmental agencies such as the Department of Defense and NASA have, consistent with policy direction in Office of Management and the Budget (OMB) Circular A-119, encouraged transition of their own standards, with appropriate input and modifications from other interested parties, into VCS. This is consistent with the increased use of such VCS by government to increase efficiency, reduce costs, and promote cooperation with commercial entities.

2.4   General Policies

Many countries have recognized the benefits of having and documenting a systematic approach to standards development and use. This is often embodied in some sort of national standards policy document(s), sometimes highly thought out and quite detailed. The typical national policy on standardization incorporates a number of elements. Some of these are listed here:

1. It notes a desire for a successful and competitive economy, and recognizes explicitly the role of standards in helping to achieve one.
2. It recognizes that no economy stands alone in the world.
3. It encourages the use of standards.
4. It identifies one or more boards, committees, or other bodies that can help guide the standardization effort, involving representation from both inside and outside the government.
5. It typically also identifies standards or categories of standards that have been or should be developed.
6. It recognizes a relationship between its own standards and international standards.
7. It addresses the issue of conformity assessment.

2.5   National versus State, Provincial, or Local Standards Involvement

While all levels of government make use of standards, develop standards, and may be involved with development of VCS, it is natural that the greatest involvement be by the governmental organizations with the largest operations and the greatest resources. The greatest standards involvement therefore takes place at the national level. Large states or other jurisdictions also play a role, however, and sometimes a very significant one because of their greater agility. An example of this is the set of rules that the California Department of Energy (DOE) is promulgating regarding hydrogen-fueled vehicles. The State of California is taking the lead, with the US DOE following. California also outlawed the (carcinogenic) gasoline additive MTBE before the federal government did.

The standards developed by these jurisdictions are referred to in various ways, sometimes falling into categories such as policies, ordinances, and regulations. The federal government has the greatest role, having a formal policy and much legislation dealing with the development and use of standards, as will be seen in the following section. Following that will be examples of lower level jurisdictional involvement.

2.6 The US Government and Standards

The government of the United States is involved with standards in all of their forms and in a number of ways. This involvement goes as far back as the writing of the constitution, which (Article I, Section 8) states, "The Congress shall have power to … fix the Standard of Weights and Measures."

The United States Office of Standard Weights and Measures was created by Congress in 1824 to establish and promote the use of uniform weights and measures. In 1836, this office was moved to the Treasury Department (which maintained standards in the customs houses for purpose of taxation). In the Treasury Department, Coast and Geodetic Survey became the responsible entity.

In 1890 the Office of Standard Weights and Measures became the Office of Construction of Weights and Measures; in 1901 its name was again changed, to the National Bureau of Standards. It moved to the Department of Commerce in 1903 with the added responsibility of promoting the use of technology as a way of assisting the international competitiveness of American businesses. In the more than one century that has since elapsed, the organization has evolved from a small office of 12 employees to a significant measurement, research, and standards developing institute with two campuses and over 3000 employees. It also went through another name change in 1988, becoming the National Institute of Standards and Technology (NIST), and along the way acquired a budget of over $800 million yearly to accomplish its many responsibilities.

In parallel with the evolution of NIST, a number of other policies and laws have been put in place to deal explicitly with the issues surrounding standards and their use. These include the Trade Agreements Act (1979), the OMB Circular A119 Federal Participation in the Development and Use of Voluntary Consensus Standards and in Conformity Assessment Activities, the National Technology Transfer and Advancement Act (NTTAA) (1995), various other laws and executive orders, and the National Science and Technology Council (NSTC) policy recommendations, in addition to numerous laws and regulations invoking the use of various specific standards.

For many years the Department of Defense was a very large developer and user of standards, having MIL or other standards for nearly anything they might purchase or use (hammers … chocolate chip cookies …). These have largely been replaced by the use of VCS and commercial off-the-shelf products.

2.7 US Government OMB Circular A119

The development of OMB Circular A-119 began in about 1977, and it was first issued in January 1980. Its purpose was to reduce the cost to the US Government by implementing VCS in lieu of the many government-unique standards then in effect – avoiding the purchase of the legendary $400 hammer. The OMB is a part of the executive branch of the US Government, and thus this was an effort undertaken by the Office of the President in an attempt to streamline government acquisitions. Since its initial issue this document has been revised a number of times, first in 1982, and then again in 1993, 1998, and 2016. While the circular was first issued by the president of the United States as instruction to agencies of the executive branch, in 1995 its principles were codified in the NTTAA of 1995, with some changes and additions. Because of the difference between the 1993 circular and the text of the law intended to implement its principles, there was a further revision of the document in 1998.

With each new revision of the circular came more explicit direction and a somewhat greater scope. In particular, the NTTAA, and the 1998 revision of OMB Circular A-119 brought into scope the concept and specific requirements for conformity assessment, an essential component of the use of standards, in addition to greater specificity regarding the use of standards themselves. OMB Circular A-119 has thus incorporated as part of its direction the ensuring of compliance with standards through conformity assessment. It has also moved from its initial role of pushing the US Government toward use of VCS, to become the implementing policy for action taken by Congress in response to that early push.

Circular A-119 thus provides direction to executive branch agencies regarding their responsibilities with respect to the use of technical standards and conformity assessment. It establishes policies on federal use and development of VCS and on conformity assessment activities and it enumerates the reasons why it does this. Reasons for the current version of this document include the following:

1. Establishing policies to improve the internal management of the executive branch with respect to the US Government's role in the development and use of standards and conformity assessment.
2. Directing agencies to use VCS rather than government-unique standards, except where inconsistent with applicable law or otherwise impractical.
3. Facilitating agencies' compliance with obligations under US trade statutes and trade agreements.
4. Providing guidance and direction for agencies participating in the work of VCS bodies and describing procedures for satisfying the reporting requirements of the NTTAA.
5. Minimizing the reliance of agencies on government-unique standards.
6. Providing policy guidance and direction to agencies on the use of conformity assessment in procurement, regulatory, and program activities.
7. Eliminating the cost to the federal government of developing its own standards and decreasing the cost of goods procured and the burden of complying with agency regulation.
8. Providing incentives and opportunities to establish standards that serve national needs, encouraging long-term growth for US enterprises, and promoting efficiency, economic competition, and trade.

9. Furthering the reliance upon private sector expertise to supply the federal government with cost-efficient goods and services.
10. Helping to meet the five fundamental strategic objectives for federal engagement in standards activities set out in Memorandum M-12-08, "Principles for Federal Engagement in Standards Activities to Address National Priorities" (http://www.whitehouse.gov/sites/).
11. Enhancing collaboration with the private sector on standards serving national needs.
12. Promoting efficiency and competition through harmonization of standards.

OMB Circular A-119 also defines what it intends by the term "standard" ("Technical Standards," or what we refer to in this volume as "Engineering Standards.") and conformity assessment. It says that consideration should be given to use of standards developed by international VCS bodies, and it requires that agencies review their standards every 5 years and replace with VCS those for which this is practicable.

2.8   National Technology Transfer and Advancement Act

Public law 104-113, the "National Technology Transfer and Advancement Act of 1995" codified policies regarding the use of standards with the goal of advancing the transfer and use of technology. It codified the OMB Circular A119 of 1993, with some changes and additions, resulting in the 1998 revision to the circular. A significant element of this legislation was the recognition that the benefit of standards is only as good as their implementation, and that an effective conformity assessment program was needed in order to ensure that the selected standards are met.

In addition to providing for transfer to the private sector of rights to inventions paid for by the government, this Act includes the following provisions:

1. NIST to coordinate federal, state, and local technical standards activities and conformity assessment activities with private sector technical standards activities and conformity assessment activities, with the goal of eliminating unnecessary duplication and complexity in the development and promulgation of conformity assessment requirements and measures.
2. " … All Federal agencies and departments shall use technical standards that are developed or adopted by voluntary consensus standards bodies, using such technical standards as a means to carry out policy objectives or activities determined by the agencies and departments [with exceptions for impracticalities and inconsistencies with applicable law]."
3. In order to meet the requirements in (2) above, "Federal agencies and departments shall consult with voluntary, private sector, consensus standards bodies

and shall, when such participation is in the public interest and is compatible with agency and departmental missions, authorities, priorities, and budget resources, participate with such bodies in the development of technical standards."

4. It requires annual reporting of exceptions to the use of VCS to the OMB through NIST.

5. DEFINITION OF TECHNICAL STANDARDS – As used in this subsection, the term "technical standards" means performance-based or design-specific technical specifications and related management systems practices.

Together the NTTAA and OMB Circular A-119 have had a significant effect on standards development in the United States. This is in spite of the fact that the Circular is directed particularly at the executive branch of the US Government. The active role of over 9000 civil servants in development of VCS, the push by the US Government to make use of VCS, the definition of what is required for a standard to be considered to belong to that category, and the identification of the roles of the American National Standards Institute (ANSI) and the NIST provide a strong incentive for use of a rigorous process to develop many more standards than might otherwise have occurred.

2.9   National Science and Technology Council

In 1993, President Bill Clinton established the cabinet level NSTC by executive order as a means of coordinating science and technology policy. The NSTC undoubtedly had a significant role in shaping the revised text of OMB Circular A-119, having in its recommendations recognized the government as a contributor to a private sector led VCS process, outlined objectives for governmental engagement in standardization activities to support national priorities, and laid out key principles for voluntary standardization processes.

2.10   Other US Government Actions

Numerous other federal actions have been taken to enhance, guide, and protect the standards creating process, including the Trade Agreements Act of 1979, a number of additional executive orders, and many specific laws and regulations invoking the use of VCS to achieve specific technical objectives.

2.11   How the Government Uses Standards

The government makes use of VCS in a variety of ways. These include the following:

1. Direct use in the programs and operations of agencies, such as use of standards for testing in the process of performing research regarding material properties or to assess the compliance of food products with limits on pesticides.
2. Writing them into contracts, agreements, or grants. An example of this is a standard for construction of a water chiller used in a building HVAC system or a standard for asphalt concrete to be installed in a paving project.
3. Incorporation of standards by reference in regulations. The Code of Federal Regulations has over 10,000 citations of VCS. The heaviest user is the Department of Homeland Security (DHS), with over 2000 citations, but the Environmental Protection Agency (EPA), Department of Transportation (DOT), Health and Human Services (HHS), Department of Labor (DOL), and others have many citations as well. The EPA would use standards for testing and for purity of products. DOL has many citations of standards in the Occupational Safety and Health Administration regulations, such as requirements for use of the American Society of Mechanical Engineers (ASME) Boiler and Pressure Vessel Code for pressure vessels containing air, oxygen, hydrogen, and various other fluids, and for compliance of head protection with various ANSI standards.
4. *Deference*. Sometimes, a VCS is successful and accepted by industry, and its use dominates a particular area. In such a case a decision might be made not to regulate that area simply because there is no need. Such a decision is most likely not made expressly and is not likely to be documented as to its basis, being in a sense a lack of a decision *to* regulate, rather than a deliberate decision *not to* do so.

2.12 US Government as a Participant in VCS Activities

While there was undoubtedly involvement of government employees in the development of VCS prior to issuance of OMB Circular A-119, the publication and evolution of an expressed policy of government participation in VCS activities has increased this significantly. This involvement helps to ensure that government needs are met; it contributes to the effort a level of expertise and resources that are sometimes not otherwise available, and it provides further benefits to the government through the exchange of knowledge and experience with industry experts. OMB Circular A-119 provides direction for employees in the executive branch of the US Government regarding this participation as well as the expected benefits. An example of this participation is the active role played by the Nuclear Regulatory Commission (NRC) in the development and maintenance of the ASME Boiler and Pressure Vessel Code Section III nuclear standards. This active approach to use of VCS also influences industry to adopt these standards as a means of meeting the nation's need for a safe and reliable nuclear industry.

2.13   State and Local Standards Use

Some states rely almost entirely on VCS, but nearly all have developed a set of amendments that they apply to some standards, particularly the building codes. Here are a few examples.

The State of California has developed a fairly extensive set of amendments to the International Code Council (ICC) International Building Code (IBC). They include two volumes of the California Building Code (CBC), a California Residential Code, California Electrical Code, and a number of others. Even the City of Los Angeles has taken a part, with the City of Los Angeles Building Code, close to 100 pages based on the CBC and the IBC. Los Angeles also has codes in specialties such as Green Building, Electrical, Mechanical, etc.

A similar situation exists with the State of New York and New York City.

Saint Louis County, Missouri, a smaller and less populated jurisdiction, deals with this issue less aggressively, specifying various VCS such as the ICC IBC, the NFPA National Electric Code, and the ASME Safety Code for Elevators and Escalators, with a few engineering requirements that delete, modify, or amplify upon specific IBC or other VCS requirements, including some differences regarding stair rails, guards on retaining walls, and construction in flood zones.

Smaller jurisdictions often adopt either the IBC or the code of the next jurisdiction up. The City of San Mateo, California, a city of approximately 90,000 on the San Francisco peninsula, has adopted the California Building Code and various other California codes for construction, which then reference, with modifications, the IBC, etc.

2.14   Other Countries

The European Union, Japan, and most other industrial countries have involvement with standards that differ only in the details from the examples given above for the United States

Under the Ministry of Economy, Trade and Industry, the Japanese Industrial Standards (JIS) Committee Secretariat has published Japan's Standardization Policy 2013, the Contents of which includes as topics JIS, Conformity Assessment, Approaches to International Standardization, International Cooperation, Human Resources Development, etc.

The European Union is involved extensively in the development of standards, including both EN and ISO standards. As a relatively recent and still evolving political entity, the EU seems to have a heightened awareness of the various issues regarding international standards (such as the ISO standards), regional standards such as the EN standards that it develops, and VCS.

2.15   Summary

Ever since the beginning of the industrial revolution standards have played an increasing role in society. Their development and use have continued to accelerate. Governments have recognized the importance of standards to economic success and advancement, and as a result the government of nearly every nation that is industrialized or hopes to become industrialized has taken a role in the production and application of standards. This typically takes the form of a formal standards policy that provides a framework for involvement of the government and its employees in standards development, as well as laws of some sort that invoke the use of certain standards.

Expressed policies on standards may be expected to increase the coherence of standards development and application, but irrespective of whether such a policy exists in a jurisdiction, standards are widely used by all governments and at all levels. One has only to look to see the wide application of standards throughout government and society.

2.16   Case Studies

Case Study 1

Consider the case of a small nation whose exports have hitherto been limited mainly to textiles and clothing products. The government of this nation believes that in order to have sufficient resources to provide for the health and stability of its population it must advance economically by developing an industrial capacity, starting with, for example, automotive electronics.

What sort of basic standards policy might it adopt?

Case Study 2

A major automobile manufacturer wishes to develop and produce a self-driving vehicle for use on public roads in the United States. Describe a framework for its use of standards, including interactions with local, state, and federal governmental agencies.

What sorts of standards will it need to be involved with, and how will it convince regulators of the safety of its products?

Case Study 3

A manufacturer intends producing a product that must meet a governmental agency's standard published many years ago. This governmental standard references a recognized international standard published at that time. However, at the present time this international standard has been updated numerous times and is clearly not the same as the edition referenced by the governmental standard.

What should the manufacturer do to proceed with producing the product?

3

Voluntary Consensus Standards and Codes

"No problems with the wheel, but this
fire thing is going to be a tough sell
to the Consumer Product Safety Commission."

(Courtesy of Cartoonstock, www.cartoonstock.com)

Primer on Engineering Standards: Expanded Textbook Edition, First Edition.
Owen R. Greulich and Maan H. Jawad.
© 2018, The American Society of Mechanical Engineers (ASME), 2 Park Avenue, New York, NY, 10016, USA (www.asme.org). Published 2018 by John Wiley & Sons Ltd.

3.1   Purpose of Standards

Engineering standards are normally sets of rules, procedures, and/or requirements developed and approved by consensus to assure the adequacy of a given
product, process, or system. Standards are normally developed in committees
whose members have experience in a particular field of endeavor. Generally,
standards are written for the following purposes:

1. Safety and reliability
2. Reduction of cost
3. Increased flexibility
4. Promotion of business
5. Helping society function
6. Consistency.

3.1.1   Standards for Safety and Reliability

Much has been learned about how to create effective standards for safety and
reliability since the early efforts. Even the codes for protection from fire have
evolved and expanded. Much of Baltimore burned in 1904 because fire crews sent
from other cities to help found that their hoses could not be connected to the Baltimore fire hydrants. This lesson was learned again in 1991 in Oakland, California,
when thousands of houses were burned, in part because fire departments from
neighboring jurisdictions found themselves unable to connect with the hydrants
in that city (Note that in neither of these two cases was this a failure of standards
due to lack of conformity assessment. In each case the city had selected a product
that happened not to be compatible with tools available in other jurisdictions.).
Standards for interchangeability thus play a significant role as safety standards.

The National Fire Protection Association (NFPA) has developed several hun-
dred standards to enhance safety. Some of these are aimed directly at preventing fires, but these standards have evolved to serve other purposes as well. The
National Electric Code (NEC) was developed by the NFPA because of the large
number of home fires caused by arcing and overheating due to overloaded wiring
and unsafe installations. It has evolved, however, to include other aspects of
safety, such as requirements for ground fault circuit interrupters to reduce the risk
of electrocution. With adoption by most jurisdictions in the United States, it has
become the accepted standard guiding the design, installation, and inspection of
practically all home and commercial electrical installations in the country. While
the primary purpose of the NEC was originally increased safety, this standard
now has the added benefits of reducing the costs of construction and insurance,
and increasing the confidence of home purchasers.

Another organization that has developed a large number of standards ($\sim$600
as of this writing) is the American Society of Mechanical Engineers (ASME). In

1883, an ASME committee on standards and gauges was created, and in 1884 the ASME published its first standard to provide uniform test requirements for boilers. ASME is perhaps most known for the Boiler and Pressure Vessel Code (BPVC), first published in 1914, but it also has standards for materials, fasteners, cranes and other lifting devices, and more. Since that time the BPVC has grown from 114 pages to comprise 28 volumes, including 12 volumes on nuclear components. The ASME B31 piping codes perform a similar role for piping systems in many applications from gas distribution to industrial refrigeration to chemical plants and petroleum refineries.

The BPVC is effective in enhancing safety because it provides a structured way of designing and constructing pressure vessels, including standard materials and designs, standard design processes, limitations on material stress, as well as nomenclature and symbology, and its use involves a rigorous conformity assessment process. Among its standards are those for materials and materials testing, ensuring predictability of material properties. These lead to greater understanding and confidence on the part of engineers using these standards, as well as ensuring designs that effectively serve their intended functions.

The BPVC volumes on welding and nondestructive evaluation help round out the BPVC, ensuring that implementation of a quality design is performed with proper process control and inspection, resulting in a product that effectively meets and safely performs its intended function.

All volumes of the BPVC are quite prescriptive, and while many choices are still made by the designer, once a design approach and materials are selected, there are many specific requirements that must be met.

Standards from the American Institute of Aeronautics and Astronautics (AIAA), reflecting the more recent development of the aerospace industry and the continued evolution in products used, tend to be less developed, and therefore more performance based, but they still aim to provide for a safe and reliable product. Because weight is of critical importance, aerospace standards provide for increased use of composites and new design concepts, but they still specify safety factors on stress and on life, as well as stringent control of quality. When new challenges are recognized, such as the previously unknown time-dependent failure mode referred to as stress rupture, standards developing committees move to update the standards in order to address them.

Finally, a simple standard of uniformity for the purposes of safety, which most of society takes for granted, is the PRNDL (Park-Reverse-Neutral-Drive-Low). This is the federally mandated order for features in an automobile automatic transmission. Because of the PRNDL we can get in a car, start the engine, and drive away, confident that we will not accidentally shift into Reverse when we intend to put the car in Low or Drive.

The importance of the PRNDL never fully impressed one driver until he had the opportunity to drive a Model T Ford. On a Model T, there are three pedals (just like in a current model manual shift vehicle – almost) and there are two levers on the steering column (again like a current model vehicle – almost). This familiar appearance gives the driver of cars with a stick shift immediate confidence, until a closer look reveals that the lever on the left side of the steering column does not operate the turn signal and high beams (there are none – no standard required them), and the right lever does not operate the windshield wipers and washers (none of those, either – who thought of them?) (That lever on the left is the spark advance–retard, and the one on the right is the throttle (no gas pedal!?). The right pedal is the brake, the middle pedal is forward–reverse, and the left pedal is high, neutral, and low. This particular driver found himself pushing the Model T away from his rental car after a fortunately very low speed collision.

Other more recent standards for safety and reliability include those for biomedical devices, computer hardware and software, and protection of the environment.

3.1.2   Standards to Reduce Cost

Many standards have been developed to provide predictability and dependability of design through use of standardized components and standard configurations. This provides for a simplified design process, permits development of preapproved lists of products and product standards, and avoids the need for requalification for every application.

Component standardization makes possible the specification of pipe sizes, for example, with the designer having confidence that the product will be available, that standard off-the-shelf fittings and flanges can easily be procured for it, that the pipe has a known pressure–temperature capability (given a particular material and wall thickness), and that future modifications will find compatible products as needed. Many of these standards include pressure/temperature ratings as well as the physical characteristics and material properties, eliminating the need for analysis every time that a component is used.

Similarly, standard door, latch, and lock sizes simplify the design process for architects and make it easy for homeowners to purchase hardware replacements and upgrades for their homes. Other standards for such things as wall electrical receptacles, tire sizes, bearings and bushings, and USB ports expand this benefit.

An engineer recently arrived from a less developed country was doing some design work. Her supervisor stopped by her desk one morning to check on progress, and was told that she was designing a bearing for a wind tunnel access hatch. When showed a catalog of standard bearings, she was almost overcome. "You can just order these!?" Because of lack of standards and lack of stock, she was used to designing from scratch essentially every component that she used. She was overjoyed to find that she could now look up standard components, specify them, and apply the standard dimensions to her designs.

The full benefit of the economies of scale that came with the industrial revolution could not be realized until standardization of components allowed manufacture of product components without the producer knowing the end product. While there will always be a need for specialized fasteners, the existence of standard screws, bolts, and nuts, available at the local auto parts or hardware store, has reduced costs immensely.

Relatively recent engineering standards include many related to information technology. Printers, scanners, monitors, memory devices, and other components can simply be plugged into the backs of computers, using USB or other plug configurations that are all available at electronics stores. Inkjet printers have become so inexpensive that companies now sell them at little more than the cost of an ink cartridge, just to get the business of selling the replacement cartridges (remember the safety razor?). Even the internal components of a computer have become sufficiently standardized that a novice, after minimal study, can purchase a housing, power supply, motherboard, memory, and other components, and with a reasonable investment of time and effort, produce a working computer capable of running standard operating systems, software applications, and games, often better than the products available on the open market as completed assemblies.

Motors with standard power, torque, and speed ratings are available, with standard mounting configurations, shaft sizes, and power demands. If a manufacturer finds a shortage of a drive component, or a user finds that a component has ceased to work, another can likely be substituted with no decrease in performance and essentially no downtime.

The same standard materials that enhance safety through predictable properties allow designers to work with confidence, without testing materials for themselves. Many materials, after having been subject to rigorous process control, have been tested and certified by the manufacturer (steel mill, plastic manufacturer, etc.) in accordance with accepted industry standards for material chemical and physical properties. This process is referred to as conformity assessment and is discussed in Chapter 8.

3.1.3   Standards for Increased Flexibility

Standards developed for increased flexibility and those for reduced cost are often difficult to distinguish, since their roles go hand in hand. Until the 1990s, students wishing to take the train from France to Madrid found that an unexpected drawback to taking the overnight train was that they were awakened at the border in the middle of the night, not just to show their passports, but because the train cars were lifted onto new trucks, or the gauge of the trucks on the cars was adjusted, to accommodate the different rail gauge then in use in Spain. It would be hard to say whether the inability to share rolling stock, the labor to change the trucks, the inability simply to keep the trains rolling, or the inconvenience was the greatest cost. In the early days of railroads in North America rail gauges were not standardized. Because of the economic challenges associated with varying rail gauges, Canada had converted to "standard gauge" by 1880, and by 1886 the railroads in the United States had converted as well.

In some cases, manufacturers have maintained internal standards different from each other or even from an industry standard. This is usually done in hopes of continuing to control a market. There are benefits and drawbacks of this approach. It is a difficult situation to maintain, and most companies that have tried have not been successful. In some cases, the standard is maintained as a secret, as with a large soft drink company, which has maintained a prominent position in the soft drink market for well over 100 years. A prominent computer company has been successful in maintaining its own standard, but it nearly went under because of it during the 1990s. Several manufacturers of tube fittings and other piping components have been fairly successful in maintaining unpublished internal standards due to a reputation for quality, combined in some cases with a certain amount of mystique, but the tendency is for these suppliers to be marginalized as more and more customers opt for the greater level of flexibility that is provided by common industry standards. The ability to purchase with confidence a product, sight unseen, is hard to beat.

3.1.4   Standards for Promotion of Business

The reader has probably by this time observed that the benefits of a single standard often fall into a number of areas. What makes a product safer, more reliable, less expensive, and more flexible in its application will typically promote product sales as well. Automobiles are more reliable because the materials of which they are made are well defined, the tolerances are good, and electronic components have been produced using precise process control. Customers who used to expect to have spark plugs replaced every three to five thousand miles now may not have to replace spark plugs during the life of their vehicle.

The same customer is also more willing to make other purchases because she knows how much light to expect from a light bulb, that the battery she orders online will work in her smart phone, and that the sheets she purchases will fit her bed (The standards for clothing sizes are less well defined, and thereby less effective.).

In years past, Swedish knives had a reputation for quality, although for a long time, there was little understanding of why. Now that standards for steel call out how much carbon, sulfur, molybdenum, and other alloying elements and impurities they contain, and how they are worked and heat treated, customers can purchase quality knives from many more sources.

In business, engineering, and industrial settings, similar motivations apply. Consistent expectations allowed by standards drive sales of machine tools, allow faster and easier design and production of heat exchangers, and help chemical producers sell great volumes of their products.

Many businesses, in order to be more competitive in a world market, adopt not only local standards but also international standards such as ISO to enable them to compete on the same level with other companies. Unfortunately, there are sometimes competing international standards. Thus, the international standard used in one continent, say Europe, may not be the same as that used on another continent, say Asia. This sometimes forces businesses doing international work to be familiar with multiple international standards for the same product.

3.1.5   Standards to Help Society to Function

It used to be that the use of credit cards and their billing entailed thousands of keypunch operators keying in millions of credit card numbers and dollar amounts every day. Engineering standards for credit and bank cards and the machines that accept them ensure that consumers can drive into a gas station almost anywhere, purchase a tank of gas, and drive off, usually without even the need to interact with the gas station attendant. The use of barcode scanning and a credit card swipe at a department or hardware store allows customers to purchase, and stores to sell, in much less time than used to be required, including (for better or for worse) a complete record of purchase, time, and other information.

Everyone knows what a stop light means, and the standardization of red light on top ensures that even people who suffer from red–green color blindness know when to stop. Similarly, a double yellow stripe on a pavement or highway indicates a no-passing zone.

Mobile phones can be purchased new or used, online or in person, and can be activated and used on the systems of multiple carriers, and the apps available for smart phones are endless.

The standards for these items, while also featuring one or more of the benefits noted in the previous sections, simply make life easier and more enjoyable, allow the traffic to flow, and help a modern society function. Similarly, standards give people the ability to purchase components with confidence, sight unseen. Hence a purchaser buying over the Internet a class 150 flange from Supplier A or Supplier B is assured of having the same quality and performance from either supplier.

3.1.6   Consistency

Standards help maintain consistency in a wide range of commercial and con-sumer products. One example is the shape and size of an electrical outlet in a given country. Many manufacturers produce electric receptacles for a given country and they all fit the appliances connected to them in that country. Of course if a single standard were used worldwide, then the receptacles and plugs would also be compatible from country to country.

Another commercial product line includes standard flanges used for pipes and pressure vessels. In the United States most of these are specified to be in compli-ance with the ANSI B16.5 standard so they can be purchased from manufacturers anywhere and still be interchangeable.

While electrical receptacles and flanges are manufactured to dimensional stan-dards for consistency, other aspects of consistency in standards involve design methodologies. Hence, a designer in the United States using the ASME BPVC will come up with the same required thickness for a given cylindrical shell as a designer using the same code in India. Such consistency in design helps world trade.

3.2   Voluntary Consensus Standards

Voluntary consensus engineering standards (often referred to as voluntary con-sensus standards, or VCS) are normally developed by committee members and then approved by members belonging to diverse interest groups that have the same level of interest in the specific standard being developed as discussed in Section 1.2.4. While the word consensus may mean different things to different people, specific procedures (standards) have been established in the United States that establish guidelines for consensus approval of standards. In the United States these guidelines flow from the National Technology Transfer and Advancement Act and compliance is ensured by the American National Standards Institute (ANSI).

3.3   American National Standard Institute (ANSI)

ANSI is a private nonprofit organization founded in 1918 by five engineering societies and three governmental agencies. Part of their published mission is as follows:

> *ANSI facilitates the development of American National Standards (ANS) by accrediting the procedures of standards developing orga-nizations (SDOs). These groups work cooperatively to develop vol-untary consensus standards. Accreditation by ANSI signifies that the procedures used by the standards body in connection with the*

development of American National Standards meet the Institute's essential requirements for openness, balance, consensus and due process.

There are at least 1100 organizations that are accredited by ANSI for writing VCS. The requirements, as published by ANSI, are the following:

In order to maintain ANSI accreditation, standards developers are required to consistently adhere to a set of requirements or procedures known as the 'ANSI Essential Requirements,' that govern the consensus development process. Due process is the key to ensuring that ANSs are developed in an environment that is equitable, accessible and responsive to the requirements of various stakeholders. The open and fair ANS process ensures that all interested and affected parties have an opportunity to participate in a standard's development. It also serves and protects the public interest since standards developers accredited by ANSI must meet the Institute's requirements for openness, balance, consensus and other due process safeguards.

ANSI is the only accreditor of organizations developing American National Standards. The Standards Council of Canada, in addition to promoting generally the use of voluntary standards, accredits standards developing organizations and approves Canadian standards as National Standards of Canada, while the International Organization for Standardization (ISO) accredits SDOs internationally. Some of the requirements that must be met in publishing VCS in the United States are published by ANSI. A short excerpt of these published requirements:

In its role as the only accreditor of US voluntary consensus standards developing organizations, ANSI helps to ensure the integrity of the standards developers that use our ANSI Essential Requirements: Due process requirements for American National Standards. A separate process, based on the same principles, determines whether standards meet the necessary criteria to be approved as American National Standards. Our process for approval of these standards (currently numbering approximately 10,000) is intended to verify that the principles of openness and due process have been followed and that a consensus of all interested stakeholder groups has been reached.

The hallmarks of this process include the following:

• Consensus must be reached by representatives from materially affected and interested parties.

- Standards are required to undergo public reviews during which any member of the public may submit comments.
- Comments from the consensus body and public review commenters must be responded to in good faith.
- An appeals process is required.

Some of the organizations that write consensus standards in accordance with the ANSI requirements are listed in Appendix B.

3.4   Codes

3.4.1   National Codes

The terms code and standard are often interchanged in conversation. However, a code is also often a body of standards grouped together for ease of reference. Technical codes cover a wide range of disciplines ranging from electrical to mechanical to structural. They form the basis for safe, consistent, and effective designs in various fields. Codes are normally specified in the design documents of a project.

There are instances where either of two well-recognized codes can be used in solving a given engineering problem and the engineer has to decide which one to use. A case in point is the National Board Inspection Code (NBIC) and the American Petroleum Institute (API) 579 code, both of which address the repairs and alterations of pressure vessels. These two codes have different requirements but the end result is close to the same for a majority of cases. An engineer repairing a pressure vessel needs to be familiar with both of these codes in order to make a rational decision on which one to use.

3.4.2   International Codes

In addition to the United States, many other countries have established codes that are used nationally and internationally. These codes are written by committees assembled by the appropriate jurisdiction. These committees operate in a manner more or less similar to those in the United States that follow the ANS process. Appendix B lists some American and international organizations that are engaged in writing standards and codes.

3.5   Some ANSI Accredited Organizations

The following sections provide a few examples of prominent ANSI accredited standards developing organizations, with a brief discussion of their products.

3.5.1   American Society of Mechanical Engineers (ASME)

ASME is one of the most prolific standards developing organizations in the United States, having developed thousands of general consensus standards, research publications, technical journals, and engineering books and pamphlets. ASME is ANSI accredited, and it relies on a large number of volunteers and numerous staff members for maintaining its activities. The background of the volunteers spans a wide range of disciplines such as engineering, metallurgy, nondestructive examination, welding, and materials. A visit to the ASME website at www.asme.org reveals a wide array of technical books and pamphlets available to the engineer. ASME also publishes, at regular intervals, over two dozen technical journals, all written by volunteers, with a small production staff that takes care of the mechanics of publishing.

A related entity, the ASME Standards Technology, LLC (stllc.asme.org), manages theoretical and experimental research related to mechanical engineering in the areas of analysis, metallurgy, welding, and nondestructive examination. Published reports of the research activities are available from ASME.

Some of the fields in which ASME develops standards:

- Boilers, pressure vessels, and nuclear components
- Elevators, escalators, and moving walkways
- Piping and pressure relief devices
- Flanges, gaskets, valves, and fittings
- Cableways, cranes, derricks, and hoists
- Screws, threads, and bolts
- Pumps.

Each of the above areas is addressed by numerous standards and codes. For example, the ASME Boiler and Pressure Vessel Code consists of the following sections:

- Section I. Rules for Construction of Power Boilers
- Section II. Materials
    Part A. Ferrous Material Specifications
    Part B. Nonferrous Material Specifications
    Part C. Specifications for Welding Rods, Electrodes, and Filler Materials
    Part D. Properties
- Section III. Rules for Construction of Nuclear Facility Components
    Subsection NCA. General Requirements for Division 1 and Division 2
    o Division 1
      - Subsection NB. Class 1 Components
      - Subsection NC. Class 2 Components
      - Subsection ND. Class 3 Components

- • Subsection NE. Class MC Components
- • Subsection NF. Supports
- • Subsection NG. Core Support Structures
- • Subsection NH. Class 1 Components in Elevated Temperature Service
- • Appendices
  - ○ Division 2. Code for Concrete Containment
  - ○ Division 3. Containments for Transportation and Storage of Spent Nuclear Fuel and High Level Radioactive Material and Waste
- • Section IV. Rules for Construction of Heating Boilers
- • Section V. Nondestructive Examination
- • Section VI. Recommended Rules for the Care and Operation of Heating Boilers
- • Section VII. Recommended Guidelines for the Care of Power Boilers
- • Section VIII. Rules for Construction of Pressure Vessels
  - ○ Division 1: Rules for Construction of Pressure Vessels
  - ○ Division 2. Alternative Rules
  - ○ Division 3. Alternative Rules for Construction of High Pressure Vessels
- • Section IX. Welding and Brazing Qualifications
- • Section X. Fiber-Reinforced Plastic Pressure Vessels
- • Section XI. Rules for Inservice Inspection of Nuclear Power Plant Components
- • Section XII. Rules for Construction and Continued Service of Transport Tanks
- • Section XIII. Rules for Overpressure Protection (in course of preparation)

The above 28 volumes contain a wealth of information needed in order to design and construct boilers, pressure vessels, and nuclear components. Other examples of ASME standards include the following:

- • A17.2. Guide for Inspection of Elevators, Escalators, and Moving Walks
- • B16.5. Pipe Flanges and Flanged Fittings
- • B30.16. Overhead Hoists (Underhung)
- • B31.1. Power Piping
- • B31.3. Process Piping
- • Y14.100. Engineering Drawing Practices.

ASME uses a wide range of member interest categories, including the following diverse groups, as part of its consensus process in developing standards. These groups consist of local as well as international representatives.

a. Fabricators and manufacturers
b. Users and operators
c. Governmental agencies and jurisdictions (such as the Nuclear Regulatory Commission (NRC) and the Coast Guard as well as state and local jurisdictions)

d. Technical consultants
e. Research organizations (such as Oak Ridge National Lab and various universities)
f. Insurance/Inspection agencies
g. Engineering and construction firms.

Personnel from the above seven categories are involved in writing various facets of the boiler code. Hence, standards written for boiler construction as well as standards written for nondestructive examination methods have to be approved by members representing the seven groups listed above.

ASME also performs a unique service by providing a Conformity Assessment process for use in industry (see Chapter 8). The process steps are generally as follows:

- ASME publishes boiler and pressure vessel standards used by manufacturers to fabricate equipment.
- ASME certifies users of these standards to ensure that they are capable of manufacturing products that meet those standards.
- The certification process consists of the following general steps:
    o An ASME authorized team visits the manufacturer's site and audits its quality assurance procedures and implementation, which address items such as
        • material traceability
        • internal inspection procedures
        • third party inspection provisions
        • hold points
        • documentation such as design report, NDE records, and material certification.
    o The team discusses its findings with an authorized representative of the manufacturer and then files its report with ASME.
    o The auditing team is usually assembled by ASME or an institution authorized by ASME such as the National Board of Boiler and Pressure Vessel Inspectors.
    o If the auditing team finds the manufacturer to be in compliance with all of the ASME requirements for a given standard, then ASME provides it with a Stamp. The stamp provided to the certified manufacturer is affixed onto their products to indicate the manufacturer's certification that a product was manufactured according to the particular standard.
    o If the auditing team finds the manufacturer not in compliance with the ASME requirements, then it files a report with ASME identifying items that are noncompliant.
    o An ASME committee (Committee on Boiler and Pressure Vessel Conformity Assessment) consisting of members from a wide sector of

the industry meets to discuss the nonconformance. A representative of the manufacturer where the nonconformance occurred is given due process by being invited to the meeting to present the user's point of view. Based on its findings, the committee then decides whether or not to grant a stamp to the manufacturer.

The above conformity process has been in use by ASME for many decades and has proved very effective in the industry. ASME cannot, however, force any manufacturer, inspector, or installer to follow the ASME standards. Unless required by jurisdictional standards (laws, regulations, local building codes, etc.), their use is voluntary.

3.5.2   American Society for Testing and Materials (ASTM)

ASTM is another organization that is accredited by ANSI for writing standards. Like ASME, it maintains a large number of standards covering a wide range of applications. However, unlike ASME, which focuses on mechanical systems, ASTM focuses more on the wide range of testing needs that affect our life. The ASTM sections are as follows:

- Section 1 – Iron and Steel products
- Section 2 – Nonferrous Metal Products
- Section 3 – Metals Test Methods and Analytical Procedures
- Section 4 – Construction
- Section 5 – Petroleum Products, Lubricants, and Fossil Fuels
- Section 6 – Paints, Related Coatings, and Aromatics
- Section 7 – Textiles
- Section 8 – Plastics
- Section 9 – Rubber
- Section 10 – Electrical Insulation and Electronics
- Section 11 – Water and Environmental Technology
- Section 12 – Nuclear, Solar, and Geothermal Energy
- Section 13 – Medical Devices and Services
- Section 14 – General Methods and Instrumentation
- Section 15 – General Products, Chemical Specialties, and End Use Products

The above 15 categories contain thousands of technical standards available to the engineer. Sections 1–3 of ASTM are referenced extensively in ASME. Accordingly, many volunteers are members of both ASME and ASTM in order to properly coordinate the activities of Sections 1–3 of ASTM with ASME material and testing needs.

3.5.3 American Petroleum Institute (API)

API publishes hundreds of standards related to the refinery industry. Many of these standards are also used at other facilities such as chemical plants, power plants, and other industrial facilities. Some standards of API are as follows:

- API 510 Pressure Vessel Inspection Code: In-Service Inspection, Rating, Repair, and Alteration
- API 579-1/ASME FFS-1. Fitness-For-Service
- API 620. Design and Construction of Large, Welded, Low-Pressure Storage Tanks
- API 650. Welded Tanks for Oil Storage
- API RP572. Inspection of Pressure Vessels.

3.5.4 UL (Formerly Underwriters Laboratory)

UL, best known for the UL listings that reassure the public of the safety of thousands of products in daily use, is able to provide these listings in part because it publishes over 1200 safety standards in diverse fields. UL is somewhat unique, being a private company that publishes standards using the ANSI process, including public review and comment.

3.5.5 National Board of Boiler and Pressure Vessel Inspectors (NBBI)

NBBI is a nonprofit organization consisting of the Chief Boiler Inspectors of most states in the United States as well as the provinces of Canada and many cities and municipalities in North America. They are the repository for all certificates of fabricated boiler and pressure vessels registered in the United States and Canada. They also audit and certify pressure vessel manufacturers, approve manufacturers for repairing pressure vessels and boilers, and certify pressure relief valves in their test laboratories. NBBI publishes many standards related to inspection and valves. Some of these standards are the following:

- NB-18. Pressure Relief Device Certification
- NB-23. National Board Inspection Code
- NB-235. Boilers and Water Heater Safety
- NB-535. Application of National Board T/O Certificate of Authorization to Test Pressure Relief Valves
- NB-550. Application of National Board VR Certificate of Authorization to Repair Pressure Relief Valves

3.5.6 American Society of Civil Engineers (ASCE)

ASCE publishes over 40 publications related to such topics as structures, hydraulics, water management, and soils. A few of their publications are listed:

- ASCE/SEI 07. Design Loads for Buildings and Other Structures
- ASCE/SEI 24. Flood-resistant Design and Construction
- ASCE/SEI 48. Design of Steel Transmission Pole Structures.

3.5.7 Institute of Electrical and Electronics Engineers (IEEE)

IEEE has over 3000 standards in such areas as

- Aerospace
- Antennas and Propagation
- Batteries
- Communications
- Computer Technology
- Consumer Electronics Electromagnetic Compatibility
- Electronics
- Green and Clean Technology
- Healthcare IT
- Industry Applications
- Instrumentation and Measurement
- Nanotechnology
- National Electrical Safety Code
- Nuclear Power
- Power and Energy
- Power Electronics
- Smart Grid
- Software and Systems Engineering
- Transportation
- Wired and Wireless.

Like many other major SDOs, IEEE sponsors research and seminars worldwide.

Appendix B lists some American and international standards developing organizations developing general consensus standards.

4

Limited Consensus Standards

4.1 Types of Standards

Limited consensus standards (LCS) are those produced following a process less open, and sometimes less rigorous, than the American National Standards (ANS) process discussed in Chapter 2. They are most often developed to benefit a particular company or other organization such as a trade group (e.g., American Boiler Manufacturers' Association or Heat Exchange Institute). Similar to other standards, benefits accrue in a variety of ways, as described in Chapter 1.

Governmental agency LCS are often developed and used because of needs that are not sufficiently broad in the economy to warrant the interest of voluntary consensus standards (VCS) developing organizations, or because even from agency to agency in the same area needs differ. It is, however, US Government policy, documented in OMB Circular A119 and the National Technology Transfer and Advancement Act, that VCS be used rather than internal governmental standards whenever practicable (see Chapter 2). Governmental agency LCS for internal use often follow a process similar to the ANSI process excepting that, since the requirements are intended only for internal use, they are subjected only to a review by the agency "internal public," not the public at large. Governmental agency jurisdictional LCS vary in the extent to which they incorporate elements of the ANSI process. They are sometimes written with little or no direct input from outside the agency, but if they are implementing regulations for laws, they are generally subject to a public comment and review period.

It is important to recognize the distinction between governmental standards generally and jurisdictional standards (also referred to as technical regulations). Major governmental agencies all have internal standards that define how they perform their jobs. A much smaller number of governmental agencies issue standards that are applicable to the operations of the private sector, enforceable by

Primer on Engineering Standards: Expanded Textbook Edition, First Edition.
Owen R. Greulich and Maan H. Jawad.
© 2018, The American Society of Mechanical Engineers (ASME), 2 Park Avenue, New York, NY, 10016, USA (www.asme.org). Published 2018 by John Wiley & Sons Ltd.

fines or other governmental sanctions. This latter group of standards is what this book refers to as jurisdictional standards (see Chapter 5).

Private sector LCS are sometimes subject to the same lack of general interest as are governmental agency internal standards, but the benefits associated with sales and control of a market are also often significant factors in the decision not to make an effort to turn them into VCS.

By maintaining a standard as LCS, the organization(s) receive many of the benefits of a VCS while maintaining greater control, and in some cases reducing the cost and time required to develop the standard, due to the less rigorous process (often less rigorous review). In some cases the LCS are developed by an organization to supplement a VCS, providing a greater level of detail, more explicit direction, or a more stringent level of quality than provided for in the VCS. An organization using an LCS rather than a VCS usually has greater flexibility to deviate from the standard, should the need arise, as it need not publicize the fact that a particular product is an exception to its usual compliance with a VCS, something often touted in sales literature.

LCS are frequently developed by an organization to fulfill a special need such as identification of a product line, specification of manufacturing details, or compliance with a given methodology. Examples of standards for product lines are the Taylor Forge and Bonney Forge catalogs for fittings and pipe dimensions. These companies established standards many years ago for numerous reasons, all of which still remain valid:

1. To reduce cost by having a fixed number of products.
2. To inform potential customers of the dimensions and configurations of available products to facilitate the design and ordering process.
3. So the customer can order the product online knowing he or she will receive the exact product needed with all the specified tolerances, etc.
4. By having a catalog the potential customers are inclined to order from the producer of the catalog, rather than other companies, knowing the exact product and quality they are receiving.
5. To help the company project an image of being a leader in its field.

There are, of course, thousands of LCS in the market place that cover every product imaginable from lighting fixtures to tools, machinery, and construction materials.

An example of an LCS for manufacturing details is the standard published by TEMA for tube and shell heat exchangers. It is published to help heat exchanger manufacturers maintain uniform design methodology and provide consistent fabrication tolerances and dimensions. It also serves to give confidence to customers through a recognized, proven, and credible approach to design.

Many similar standards exist in industry covering various topics such as expansion joints and refractory applications. While these LCS are developed for the benefit of a given organization, they are typically used throughout the industry.

An example of an LCS for compliance with a given methodology is NASA-STD-5006, General Fusion Welding Requirements for Aerospace Materials Used in Flight Hardware, the standard developed by NASA for welding of aerospace components. This standard is used internally at NASA, and is often cited as a requirement on contracts for flight hardware. It includes details regarding equipment, calibration, maintenance and records, materials, weld procedure, performance and welder qualifications, and many other aspects of producing a quality weld.

In some cases, LCS become industry standards, and ultimately VCS. For example, Taylor Forge sponsored research at Yale University to develop design rules for flanges. Taylor Forge then published its procedure and provided forms to simplify the design in a small flange design handbook for engineers. Professor Waters and Frank Williams, one of his students who had worked on the development effort and had been hired by Taylor Forge, were asked to join ASME and help incorporate the rules in the BPVC. The Taylor Forge design process was thereby incorporated in the ASME BPVC many years ago. Similarly, portions of MIL-STD-1522A, Standard General Requirements for Safe Design and Operation of Pressurized Missile and Space Systems, have been incorporated in the American Institute of Aeronautics and Astronautics (AIAA) S-081, Space Systems – Composite Overwrapped Pressure Vessels (COPVs) standard.

4.2   Proprietary versus Nonproprietary Standards

4.2.1   Proprietary Standards

Many organizations have proprietary standards of some sort or another. These standards include the following:

- Unique product and manufacturing processes
- Patented products and procedures
- Intellectual property.

While falling under the general category of LCS, these standards are often further limited either through patent rights or by maintenance as trade secrets.

A classic example of a standard for a unique product and manufacturing process is the recipe for Coca Cola. The company keeps a tight control on the production of the product without divulging its recipe outside the company. They chose not to patent the product in order to keep the ingredients secret. Many other companies have unique product and manufacturing processes that they do not share outside the company in order to maintain a market advantage.

Many companies have proprietary standards based on a patented technology. The patents give them protection from competition for a given number of years.

In the United States, a design patent has a term of 14 years and a utility patent 20 (17 years before a change in the law in 1995). The companies develop manufacturing standards based on these patents. They then hope that by the end of the term of the patent they will have been established as the market leader such that competition will have trouble taking a significant portion of the market, or they may view the market as one that will endure for a number of years, but that by patent expiration they will have received enough benefit from their exclusive rights in the market to have justified their initial investment. This is often the case with drugs, which typically become available as generics shortly after patent expiration.

Some companies, especially engineering companies, develop a process or product upgrade that is marketed as an improvement over existing products or operations. These companies then develop proprietary and nonproprietary standards based on such intellectual property. Sometimes, the whole worth of a company is based on the value of its intellectual property.

4.2.2   Nonproprietary Standards

Many nonproprietary LCS are published by organizations to promote their product or process as discussed above. Many such standards provide valuable information to others in the field, but it must be kept in mind that the published information may be slanted toward the interests of the specific organization.

4.3   Governmental and Jurisdictional Limited Consensus Standards

Many federal, state, and local governmental agencies as well as local jurisdictions publish nonconsensus standards to provide requirements and guidance for performing their work. Some of the more well-known agencies are mentioned in the following sections.

4.3.1   NASA

NASA maintains a library of hundreds of policy directives, procedural requirements, standards, and handbooks. Many of these deal with the administrative and other issues associated with running a large organization, but the list includes dozens of engineering standards ranging from implementing requirements for safety of pressure systems through how to assure the integrity of software developed by or on behalf of the agency. In addition, NASA throughout the years has published a wealth of technical bulletins ranging in scope from buckling to composites, stress analysis, and many other topics.

4.3.2   Army Corp of Engineers

The Corp of Engineers includes hundreds of engineering manuals in its list of publications. Their applications range from how to validate analytical chemistry laboratories to safety, environmental quality, landfill gas collection and treatments systems, design of flood control structures, calibration of soils testing equipment, and more. It also maintains a wide range of other publications, including engineering circulars, design guides, regulations, technical letters, and engineering graphics standards. These are available on its website at http://www.publications.usace.army.mil/Home.aspx.

4.3.3   National Institute of Standards and Technology (NIST)

NIST publishes numerous standards and other documents in a much wider range of fields than its name would imply. It maintains, for example, a series of Federal Information Processing Standards that range from Personal Identity Verification of Federal Employees and Contractors through Security Requirements for Cryptographic Modules, and it publishes numerous interagency or internal reports (which are, however, generally available for the use of the public) related to information security. Its list of special publications includes categories such as Calibration Services, Standard Reference Materials, and Precision Measurement and Calibration, and also includes Law Enforcement Technology, Computer Systems Technology, and others.

4.3.4   National Science Foundation (NSF)

NSF publishes hundreds of standards (called policies and procedures) in various fields such as biology, computer science, engineering, geoscience, math, and physical science. Many of these provide guidance and/or requirements for researchers performing work under NSF programs. Subtopics range from requirements for research in Antarctica to laboratory equipment.

4.3.5   US Department of Agriculture (USDA) – Forest Service

The National Forest Service maintains an extensive system of directives, consisting of manuals and handbooks. These are used to provide for internal management and control of programs and as a source of direction to Forest Service employees. They include a number of categories, one of which is engineering, which, due to the broad range of activities that take place in national forests, includes a comparably wide range of subtopics. These include items such as geotechnical engineering (which references standards from the American Association of State Highway and Transportation Officials

(AASHTO) and ASTM), aerial adventure course design and manufacture (largely performance based), passenger ropeways (ski lifts, ski tow ropes, etc.), and building design and construction. The Forest Service standards are typically implementing standards for legislation or jurisdictional standards (e.g., OSHA regulations), and therefore largely refer either to regulations or to VCS. Technical details tend to be included in the Forest Service handbooks rather than the manuals, but are still written at a generally high level.

4.3.6   United States Food and Drug Administration (FDA)

The FDA regulates products and activities including food products, drugs, medical devices, cosmetics, tobacco products, and more. The standards that it promulgates are in two categories. The detailed requirements in many areas are enshrined in VCS, and FDA takes a significant role in development of these standards, participating in many SDO activities. FDA is also a jurisdictional organization and maintains many regulations (found in Title 21 of the Code of Federal Regulations) covering these areas.

4.3.7   Municipalities

Many municipalities issue nonconsensus standards regarding compliance with their municipal ordinances. However, most of these standards consist of a compilation of names and titles of VCS that must be adhered to in order to conduct business with or within the municipality and for the safety of its citizens.

Appendix C lists some industrial organizations that publish LCS.

4.4   Case Studies

Case Study

A manufacturer develops a new type of tube serrated both inside and outside to improve heat transfer when installed in a heat exchanger.

Discuss the advantages and disadvantages of producing the tube to a nonproprietary limited consensus standard versus an international consensus standard.

5

Jurisdictional Standards

5.1 Regulations and Jurisdictional Requirements

A special category of standards that affects practically all organizations is that group of regulatory standards promulgated by federal, state, and local jurisdictions such as OSHA, DOT, DOE, EPA, Cal-OSHA, counties, cities, and regional governing bodies. These standards have the distinction of being enforceable by legal sanction, typically fines, and in some cases incarceration. This is different from standards that may have been agreed upon by contract between two companies, which do not have the same means of enforcement.

(Courtesy of CartoonStock, www.cartoonstock.com)

Primer on Engineering Standards: Expanded Textbook Edition, First Edition.
Owen R. Greulich and Maan H. Jawad.
© 2018, The American Society of Mechanical Engineers (ASME), 2 Park Avenue, New York, NY, 10016, USA (www.asme.org). Published 2018 by John Wiley & Sons Ltd.

A regulation is an implementation of a law established by an official governmental jurisdiction for the purpose of achieving specific public good. It is put in place after action by a legislative body either instructing or authorizing an agency to create regulations in a particular area. The general category of regulations includes both administrative requirements and engineering regulations, the latter of which often have specific engineering requirements, but also almost always reference various local, national, or international codes and standards. As an example, many jurisdictions, especially at the state level, require compliance with the ASME Boiler and Pressure Vessel Code for pressure vessels, the International Building Code (IBC) for general construction, and the National Electrical Code for electrical installations. The practice of giving voluntary consensus standards (VCS) the force of law by referring to them in a law or regulations is referred to as "incorporation by reference" and is dealt with in more detail in Section 5.3.

Compliance with these codes becomes a legal requirement since they are specified in the regulations.

Because jurisdictions typically have wide responsibilities, are subject to various standards of openness, and for various reasons may not be ready to adopt the latest standard, it is common to have the required code or standard edition out of phase with the most recently published edition of a given code. This is typically the case with the National Electrical Code, for example, where reference to a given edition by a jurisdiction often lags behind the current code edition by several years. The OSHA regulations typically reference the VCS in effect at the time the regulations were written. For most regulations, this means the edition of standards that were in effect in the early 1970s (although a formal OSHA Letter of Interpretation states that compliance with a later edition will be viewed as a de minimis violation if the later edition provides an equal or greater level of safety). Care must be exercised to ensure that the correct edition of the standard is followed as well as identified in any contract documents.

The following table lists some jurisdictional organizations, with a brief discussion of the standards provided by each. Following this listing is a more detailed discussion of one organization (OSHA), its regulations, and their application. This list includes examples of state occupational safety organizations along with a number of organizations at the federal level. It does not attempt to address the myriad of city, county, regional, and other lower level jurisdictions.

Organization	Jurisdiction
Cal-OSHA, VOSH, Oregon OSHA, etc. (see OSHA)	These are state organizations performing the same role within a state that OSHA performs generally. In order for a state program to be recognized by Fed OSHA as a suitable program in lieu of the federal one, the state occupational safety and health organization must demonstrate a program

Organization	Jurisdiction

essentially equivalent to the federal program. A number of states have chosen to implement such programs for a variety of reasons, but typically because they believe that they can achieve the same results with less intrusion or by means more favorable to local businesses. In some cases, they may choose to be more stringent than Fed OSHA

CPSC   The Consumer Product Safety Commission maintains many standards for product safety, ranging from architectural glazing materials through bicycle helmets, infant bath seats, and baby cribs. Many of the CPSC standards are administrative and relate to product labeling, but there are many performance and other standards for individual products

DOE   The Department of Energy is a cabinet level department that was created by the Department of Energy Organization Act of 1977, which combined the Federal Energy Administration, the Energy Research and Development Administration, and the Federal Power Commission, along with some programs from other agencies. Originally concerned only with nuclear material, DOE responsibilities have expanded to include energy conservation, domestic energy production, and a significant amount of research in genomics and other physical sciences. It operates a number of well-known national laboratories, including Argonne, Brookhaven, Lawrence Livermore, Los Alamos, Oak Ridge, Idaho, and others
DOE publishes a wide range of directives used to manage its efforts. Many of these are identified as guides and do not contain actual requirements, so they do not rise to the level of actual standards. It is also responsible for a large number of regulations relating to energy conservation, nuclear safety management, occupational radiation protection, chronic beryllium disease prevention, and worker safety and health for DOE contractors

DOT   The mission of the Department of Transportation is to "Serve the United States by ensuring a fast, safe, efficient, accessible and convenient transportation system that meets our vital national interests and enhances the quality of life of the American people, today and into the future." Because of this broad mission, the DOT has equally broad portfolios of responsibilities and regulatory authority. It includes under its

Organization	Jurisdiction
	authority the National Highway Traffic Safety Administration (NHTSA), the Federal Aviation Administration (FAA), the Federal Highway Administration (FHA), the Pipeline and Hazardous Materials Safety Administration (PHMSA), the Federal Motor Carrier Safety Administration (FMCSA), and others. It has regulations that address such issues as airline scheduling in the air traffic control system and rest requirements for truckers, safety of interstate motor carriers, and construction of roads and bridges. The regulations on truckers' logbooks are purely administrative, but there are also many regulations providing technical standards for concerns such as passenger and commercial vehicle safety devices, ground and aircraft based vessels for transporting pressurized gases, the process by which aircraft type licensing occurs, and minimum requirements for securing certain types of cargo in trucks (generally performance standards). It has teamed up with ASME to publish a national standard for cargo transport tanks (ASME Boiler and Pressure Vessel Code, Section XII, Rules for Construction and Continued Service of Transport Tanks)
EPA	The Environmental Protection Agency is a relatively small government agency (<0.5% of the nation's budget), but has a disproportionately large impact on society due to the large number and range of its regulations. Close to 10% of its budget is related to civil and criminal enforcement of regulations covering climate change and air quality, water quality, cleanups, and chemical safety and pollution prevention. As with the other agencies listed, some EPA standards are essentially administrative. The general public is perhaps most familiar with EPA vehicle emissions and mileage standards. These affect everything from design of vehicles to the type of fuel that is available at gas stations
FDA	The Food and Drug Administration covers a wide range of areas including, as its name indicates, food and drugs, but also extending to medical devices. Most of its regulations would not be considered standards, but it maintains a significant database of "recognized consensus standards" for medical devices, ranging from test methods for calibration to acceptable materials for various applications

Organization	Jurisdiction
NRC	The Nuclear Regulatory Commission was established by the Energy Reorganization Act of 1974. It is responsible for enforcement of a number of regulations regarding civilian use of nuclear material and dealing with nuclear waste. Some of these regulations are technical and can justifiably be considered engineering standards, while others fall into the classification of general laws and regulations. Because its operations relate closely to and may overlap with those of other agencies, it has Memoranda of Understanding with those agencies, including the Departments of Justice, Labor, and Transportation, and with OSHA. The NRC is heavily involved with the ASME in writing rules for nuclear power plant components (ASME Boiler and Pressure Vessel Code, Section III, Rules for Construction for Nuclear Facility Components and Section XI, Rules for Inservice Inspection of Nuclear Power Plant Components), and also the ASME NOG (Nuclear Overhead Gantry) standards for electric overhead and gantry multiple girder cranes used at nuclear facilities and components of cranes at nuclear facilities
OSHA	Created by the Occupational Safety and Health Act of 1970, the Occupational Safety and Health Administration (OSHA), a part of the Department of Labor, is charged with assuring safe and healthful working conditions for employees by setting and enforcing standards, and by providing training, outreach, education, and assistance. OSHA regulations cover most private sector employers and their workers, and many public sector employees. Regulations (standards) cover most aspects of the workplace, and for those areas not specifically addressed elsewhere, the "general duty clause" has requirements for "employment and a place of employment which are free from recognized hazards that are causing or are likely to cause death or serious physical harm." In general, formal interpretations as well as court cases have interpreted this to mean compliance with industry standards, although a responsible, effective, equivalent program is accepted. Specific areas for which standards are provided include agriculture, construction, general industry, and maritime employment. In areas where detailed standards are not provided, such as pressurized equipment, OSHA references other national standards such as the ASME Boiler and Pressure Vessel Code

5.2   Jurisdictional Standards Implementation

It is typically the responsibility of businesses, including employers, airlines, mine owners, medical device manufacturers, and others to comply with jurisdictional engineering standards. While each jurisdiction has oversight authority, the extent to which this is exercised varies. In the case of medical devices, for example, the FDA performs a review of each device before accepting it for use. Within DOT, there are various administrations that use varying types and levels of oversight. The FAA has extensive certification procedures for aircraft designs, especially so for those that will be used as passenger aircraft, while the FHA has highway construction standards whose implementation is most often by contract. OSHA has extensive standards for employee safety, and has authority to inspect workplaces, but with over 100 million workplaces subject to its regulations, OSHA inspections are, on the average, rare. OSHA and the state OSH organizations conducted a total of approximately 100,000 workplace inspections in 2004 meaning that the average workplace will be subject to an OSHA inspection about once every 1000 years. Clearly, in this case, the good faith of most employers is very much depended on, while inspections are targeted toward those organizations identified as posing the greatest threat to employee safety.

Because of the ability of many regulatory agencies to levy fines or other sanctions as well as because of the civil liability that is implied by noncompliance in the event of a mishap, there is incentive (in addition to that of doing the right thing by their employees or customers) for companies to maintain compliance. The truth, however, is that implementation of jurisdictional standards is very much up to those being regulated by them. Thus far, while there have been some notable exceptions, this approach has worked very well.

5.3   Incorporation by Reference

Incorporation by reference refers to the practice of codifying material published elsewhere by referring to it in a regulation. This custom is useful, efficient, and has significant benefits for society, but it also entails challenges.

By reviewing and accepting the requirements of a VCS as its own, a jurisdiction saves the tremendous investment of resources that would be required to develop its own standard to the same level of detail and quality. It typically joins many other jurisdictions in adopting the same standard or standards, benefiting from the experience and knowledge of many of the foremost experts in the field, while also providing consistency in requirements among cities, counties, and states. Those subject to the requirements benefit in that they can apply one familiar standard, typically recognized as industry best practice, possibly in a number of different jurisdictions. Thus, a company installing home heating systems, for example, can often expect to be required to meet the same requirements regardless of the city, county, or even state in which it is performing work. This reduces risk to the installer and at the same time provides benefits to the homeowner in reduced costs and less confusion as to what is required.

For over 40 years the use of VCS to the extent reasonably possible has been the policy of the federal government. This was first documented in OMB (Office of Management and Budget) Circular A-119, followed by the National Technology Transfer and Advancement Act of 1995. This policy has been taken to heart, and approximately 10,000 VCSs have by this means become a part of the law. States and local jurisdictions have made extensive use of the same approach, although often without an explicit policy.

The results of this policy have largely been successful, but its value and appropriateness are sometimes questioned. Criticisms typically revolve around a few points.

5.3.1 Access to Reference Standards

Incorporation by reference often means that the user of the standard does not have full access to it without purchasing a copy of the standard. The typical solution to this problem has been maintenance of a reference copy of the document in the offices of the regulatory agency that references it. This is clearly unsatisfactory in many cases. Travel to Washington, DC, to review an ASME standard in the offices of OSHA is hardly practical for a company in a faraway state. Incorporation of the actual text of the standard has been considered, but involves copyright issues as well as in increase in the length of the regulation, defeating one purpose of the policy. Posting the standard online has also been discussed, but also involves copyright issues. Changing the copyright laws to allow such publication has been considered, but any of the proposed approaches involves diminishing the value of the copyrighted material. It is usual for the SDO producing a standard to use the sales of that standard to finance many of the costs of developing it. Significantly changing the copyright laws in this area would significantly change the business model of SDOs, likely not for the better. An approach followed by the Department of Homeland Security was to purchase licenses to distribute and allow download of various ASTM standards for a flat fee.

5.3.2 Updating of Reference Standards

A second challenge, referred to briefly above, is the issue of keeping a regulation current as referenced standards evolve. It has historically been considered important to avoid allowing a regulation to be changed by an outside body (the SDO) and to avoid confusion, so it is usual to specify the edition of any referenced standards. There is even a potential constitutional issue regarding delegation of authority of or by congress if the edition is not specified.

On the other hand, it is usual that products are produced to the current, or at least a very recent, version of a VCS. This can cause products that strictly comply with a regulation to be unavailable, and the available products to be technically out of compliance. Several approaches have evolved to address this issue. Some agencies use an equivalence approach, assessing later versions of a standard for equivalence. Technical amendments to regulations are sometimes issued. At times regulatory agencies have chosen informally to ignore technical noncompliances resulting from following a different version of a standard. A more formalized approach, reducing uncertainty for the regulated parties, was taken by OSHA with the issuance of a formal letter of interpretation stating that compliance with later versions of a reference standard would be considered a de minimus violation if the standard provides a level of protection equal to or greater than that provided by the referenced version.

Various other approaches have been considered, including the use of simplified procedures for changes to regulations that are not expected to be controversial, and writing into regulations a requirement that SDOs notify the regulatory agency allowing the revision to be explicitly addressed. While not yet in widespread use, these approaches may become more popular as their success in balancing the needs for transparency, stability, and flexibility are recognized.

5.4   Sample Jurisdictional Standard: The OSHA Regulations

Jurisdictional standards can accomplish their purposes by expressing specific items that must be accomplished or they can operate by specifying results to be achieved. The OSHA regulations most often do the former, both for consistency of application and because they were mainly developed at a time (early 1970s) when the safety discipline was not yet as well developed as it is today. They tell a person or employer what to do, often in great detail. When specific requirements are given, each requirement must be complied with, whereas when a performance standard is provided it is the responsibility of the affected party to determine and implement a means of achieving the required level of performance. The OSHA regulations provide safety standards for almost all fields of employment, however, and the approach does vary. For the cases in which there is not an applicable regulation, the General Duty Clause (see Section 5.4.1) is applied.

In the section of the OSHA regulations for general industry for overhead cranes (29 CFR 1910.179), many particulars of operational limitations and required inspections are specified. For example, there is a fairly detailed discussion of how the end of a wire rope may be attached to the drum. The OSHA section on Process Safety Management (PSM) of Highly Hazardous Chemicals (29 CFR 1910.119) (introduced in 1992, 20 years after most OSHA standards), on the other hand, is somewhat an anomaly in the OSHA regulations, being essentially a performance standard. Rather than providing explicit quantity–distance limits, as is done in the related section on explosives safety

(29 CFR 1910.109), the PSM section requires the employer to perform process hazard analyses on the processes covered by the standard, providing a list of six different suitable analysis methodologies, and allowing for others as well. It lists a number of items that must be addressed in the analysis, and then requires that the employer establish a system to address the findings of the analysis, that the analysis be updated and revalidated at least every 5 years, that a system be in place to manage changes, and that employees be trained and informed throughout. The employer is allowed wide latitude in implementation, provided the approach is effective.

Regarding another example of a prescriptive requirement, an employer-user purchasing an air receiver, in order to ensure compliance with the OSHA regulations in 29 CFR 1910.169, must purchase a vessel that has been designed and constructed in accordance with the ASME Boiler and Pressure Vessel Code, Section VIII, Division 1. To meet the letter of the law, the 1968 edition (that edition is cited in the law because it was current at the time the regulation was written) of the code must be used. Flexibility to use a later edition of the code is provided by a formal letter of interpretation [11], which allows use of later editions if they provide an equal or greater level of protection, which they are recognized as doing. The user would therefore normally specify current ASME code construction in the contract with his supplier, and would obtain a vessel designed and constructed, with documentation and an ASME code stamped nameplate (referred to overall by OSHA as "Nameplate, Records, and Stamping," or NRS), in accordance with the ASME Boiler and Pressure Vessel Code. If the supplier provides a vessel that does not meet these requirements, that is a matter between the user and the supplier, but the user who accepts the noncompliant vessel and puts it into service could be cited for an OSHA violation. Another letter of interpretation addresses what to do in the case of a vessel for which NRS are not available [11].

In cases where there is confusion, ambiguity, or simply a question as to the intent and application of a regulatory standard, it is usual for the jurisdiction to accept letters of inquiry (as do other consensus standard developing bodies). These letters of inquiry are responded to with letters of interpretation, formalizing the position of the regulatory body on the issue and publishing it for use of all employers. Such letters of interpretation provide guidance both to the employer and to the inspectors and other personnel in the regulatory organization, helping ensure uniform application of the regulations.

If an employer other than a federal agency wishes to operate some aspect of its business in a manner that does not comply with the OSHA regulations, then in order to avoid the possibility of citations, fines, or other sanctions, a waiver is normally required. For a federal agency employer an Alternate Standard (submitted to, and reviewed and approved by OSHA) takes the place of a waiver. Permission to deviate from requirements, which sometimes goes by different names for different standards developing organizations, is discussed in detail in Chapter 9.

Reference standards in regulatory requirements often get out of date. The intention is not to let this happen. However, because of workload as well as the barriers to changing regulations it often does. In order to make a significant change to any of the OSHA regulations there is a fairly rigorous review process that must be followed, with the result that some changes never get proposed, and some, if proposed, never get implemented. The public review process, while important to allow stakeholders a chance to weigh in on regulatory actions, can also stifle progress.

5.4.1 OSHA General Duty Clause

Many OSHA standards are quite prescriptive, while others such as the "Process Safety Management of Highly Hazardous Chemicals," which provides performance standards in a specific area, are broadly written, allowing much flexibility to the user. When congress wrote the OSH Act of 1970, creating the Occupational Safety and Health Administration, it included a catch-all statement in Section 5(a)(1) of the act, requiring that each employer "shall furnish to each of his employees employment and a place of employment which are free from recognized hazards that are causing or are likely to cause death or serious physical harm to his employees." This requirement is generally referred to as the General Duty Clause. Most major sections of the OSHA regulations do not repeat this requirement, requiring reference to the law itself when this requirement is to be applied, but the statement is repeated verbatim in 29 CFR 1960 Basic Program Elements for Federal Employees (found in 29 CFR 1960.8(a)).

This statement is applied only to those aspects of an employment that are not covered by more specific regulations. If an employer complies with all of the specific OSHA standards applicable to its operations, then it is considered to be in compliance with the General Duty Clause with respect to that issue. Thus, in the General Industry regulations, for example, 29 CFR 1910.5(f) states, "An employer who is in compliance with any standard in this part shall be deemed to be in compliance with the requirement of 5(a)(1) of the Act, but only to the extent of the condition, practice, means, method, operation, or process covered by the standard."

Examples of categories falling under the General Duty Clause, for General Industry, are in-plant storage of pressurized nitrogen for process applications and requirements for lifting hardware not specifically addressed by 29 CFR 1910.184. In these cases, the usual OSHA approach would be to apply the most applicable industry VCS. Thus for in-plant pressurized nitrogen storage, OSHA would expect the employer to follow the ASME Boiler Vessel and Pressure Vessel Code unless the employer provides justification for some other approach to ensuring employee safety. For lifting appliances not addressed by the regulations, OSHA would expect compliance with the ASME standard for those particular devices. In either case, however, if an employer has in place a safety plan that addresses the issue effectively, it could normally be expected that OSHA would accept that plan in lieu of the ASME standard.

Also, when an employer performs operations for which the only OSHA standard is the General Duty clause (i.e., there is no specific applicable OSHA standard for the particular operation or area of operations), a very broad performance standard is being applied, and there may be an employer responsibility to develop and document its own standard. For operations of covered government agencies the requirement is specifically stated in 29 CFR 1960.18, Supplemental Standards, and the Supplementary Standard (the terms used in the title and in the text of the regulation are not the same) must meet certain review requirements and be submitted to OSHA for approval. For the private sector the responsibility is more generally included in the requirements for company safety plans, job hazard analyses, etc., but reference and enforcement directly to Section 5(a)(1) of the OSH Act may be expected if those company documents are found to be inadequate to ensure the safety of employees.

5.5   Summary

The above discussion examines some of the challenges facing organizations with jurisdictional authority, and the example illustrates some of the regulations and operations of a single such agency. Details vary from organization to organization, and from standard to standard, with a wide range of regulations and enforcement styles. The most significant reasons for organizations achieving compliance are the desire to do the right thing or a concern regarding liabilities and/or jurisdictional enforcing actions.

Appendix D lists some jurisdictional agencies.

6

Standards Development Process

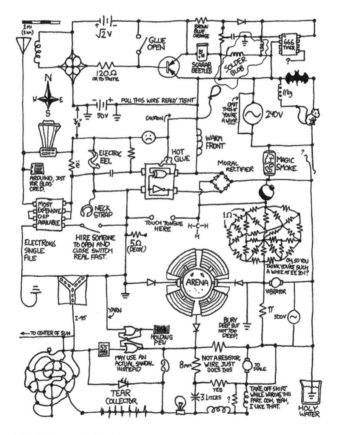

(Courtesy of XKCD, www.xkcd.com)

Primer on Engineering Standards: Expanded Textbook Edition, First Edition.
Owen R. Greulich and Maan H. Jawad.
© 2018, The American Society of Mechanical Engineers (ASME), 2 Park Avenue, New York, NY, 10016, USA (www.asme.org). Published 2018 by John Wiley & Sons Ltd.

6.1 Introduction

As discussed in Chapter 1, there are a number of reasons to develop a standard. Depending on the standard type, various processes may be followed in its creation.

If the broadest acceptance of a standard is desired, and particularly if it is thought that the standard will be appropriate as a mandatory standard to be invoked by laws or regulations, then voluntary consensus standards (VCS) are likely the most appropriate type and a fairly formalized process is required. If a law has been passed requiring that a regulatory standard be developed, the process has some features in common with VCS (e.g., public review), while the various types of corporate and other limited consensus standards typically follow their own processes.

6.2 Voluntary Consensus Standards (VCS)

The general category of VCS in the United States is defined in law (the National Technology Transfer and Advancement Act). Some very specific requirements govern the process used to develop standards that will receive this label, and the standards that meet these requirements have a greater influence as a result. The requirements generally are these: pertinence, openness, transparency, due process, and consensus. To ensure that these requirements are met, the standards developing organization (SDO) submits its process to the American National Standards Institute (ANSI), which then proceeds with both an internal and a public review of the materials and formal vote by the ANSI Executive Standards Council. Most SDOs in the United States use ANSI accredited procedures, in some form or another, to develop standards. The following procedure, used by ASME, typifies the process.

- A request is made to ASME to develop a standard or a new rule within an existing standard.
- The ASME staff assigns a tracking item to the request and forwards it to the chairman of the appropriate Committee, such as the BPV Committee on Power Boilers (Section I), within one of the Standard Committees mentioned in 2.5.1, Boiler and Pressure Vessel Committee.
- The Committee chair assigns the proposal to a working subgroup to evaluate it. The subgroup normally consists of people who specialize in a given field such as design, materials, or fabrication.
- If the evaluation supports the need to develop a standard or a new rule then the subgroup assigns personnel to work on it.
- After completion, the proposal is discussed at the subgroup level with participation from the public. Any comments or suggestions received must be discussed and voted on.
- The item then moves on to the appropriate Boiler and Pressure Vessel Standards Committee for further consideration. The personnel at the committee

level have a wider diversification in their backgrounds than personnel at the subgroup level. Discussion at the committee is also open to participation from the public.

- Following discussion at the Standards Committee, authorization is obtained to ballot the proposal as a standards action.
- A 30 day ballot is then sent to the appropriate Boiler and Pressure Vessel Committee for their approval. This ballot is also sent for review and comment by the Technical Oversight Management Committee (TOMC) and the concerned supervisory board for review of the technical content and possible conflicts, etc. For proposals pertaining to the nuclear sections of the code, the ballot is sent to the Board on Nuclear Codes and Standards. All others types of proposals for pressure system rules or standards are sent for review by the Board on Pressure Technology Codes and Standards.
- The public is given a formal review period for all proposed revisions to existing or new standards. The announcement for public review is made in the ANSI Reporter and on the ASME website. The public review period is 45 days and the proposals can be downloaded from the ASME website.
- The proposal is considered approved when two thirds of the appropriate standards committee return an "Approved" ballot and there are no more than three outstanding "Disapproved" votes.
- Due process is provided to ensure that comments, negative votes, and appeals whether from inside ASME or from the public, receive serious consideration.
- The proposal is then sent to the concerned supervisory board for a procedural review. For proposals pertaining to the nuclear sections of the code, the review is done by the Board on Nuclear Codes and Standards. All others types of pressure system rules or proposals are reviewed by the Board on Pressure Technology Codes and Standards. This review is the final step in the approval process prior to accepting the proposal as a new standard or new rule within an existing standard.

The procedure above involves "due process" in developing the standards by including a wide sector of industry in writing the rules, by providing for public comment, and by ensuring that any comments are fairly evaluated. Many other organizations follow a similar procedure and process.

Note that after following all of this process, a VCS is still just that, a *voluntary* consensus standard. Only if it is required by a regulation or other governmental action do any of its requirements become mandatory (see Chapter 2).

6.3   Government Nonjurisdictional Standards: DOD, NASA, etc.

In addition to those governmental standards that have force of law, essentially all governmental agencies produce standards used either internally or externally to aid in accomplishing their responsibilities. Those used internally are to document and give structure to policies, procedures, and processes used inside the organization, while others are used in working with other organizations, typically contractors performing work for the government.

Process details vary widely among organizations. The nature of the Department of Defense, for example, leads it to be more hierarchical and top down than some of the other organizations. The Forest Service has a document structure that has standards that apply to different regions of the country. Federal Highway Administration standards, while affecting the private sector, also come with money when they are implemented on a contract, so a less rigorous process may not be objected to by those affected. NASA follows a process so inclusive and rigorous that, except for being internal to the organization, it very nearly mimics the American National Standards (ANS) process.

With this variability in process, it is to be expected that there will be a range of qualities among governmental nonjurisdictional standards, and this is indeed the case.

6.4   Governmental Jurisdictional Standards: DOT, FAA, FCC, OSHA, etc.

Governmental jurisdictional standards, sometimes referred to as technical regulations, come about because of legislation either authorizing or directing a governmental body to produce requirements in a particular area. If specific requirements are put directly into place by legislation, then there is no formal public review (although the court of public opinion can be strong). In the case of technical regulations there are requirements for public review that can in some cases be as strict as those in the ANS process. This helps minimize the number of deficient regulations that actually take effect, as well as ensuring that those most directly affected by a standard have a say in its writing.

6.5   Corporate Standards

Most large corporations have internal standards used for daily operations. These standards are typically categorized proprietary. They often have another set of standards that, while not reviewed by the public, are made available to them.

6.5.1   Corporate Public Standards

Public standards are used by corporations to inform the public and their customer base of information related to their product as discussed in Section 3.2. There are literally millions of such standards in both printed and electronic formats. These standards range in topics from computers and computer chips to pipes and fittings. Many of these company public standards become national standards as time goes on or are incorporated into existing standards. The standards are developed over some time and they may start with one product or service and then slowly increase in volume to cover a wider range of needs. Many of these standards, while not created or maintained using the ANS process, are intended to provide information to assist the public in using or specifying company products, so they are made publicly available and can often be accessed over the internet. A few private organizations, such as UL, follow the rigorous ANS process to produce standards that are used in much of their work and as standards for the public safety as well.

6.5.2 Corporate Proprietary Standards

All corporations have proprietary standards used for their operation as discussed in Section 4.2. Their development is necessitated by the need to operate consistently or deal with products in a systematic manner. These standards are typically kept proprietary in order to give the corporation an advantage over competitors, although it is sometimes a matter of simply not wanting them to be open to question by outside entities. In many cases these standards are purposely not patented for competitive reasons. This is because once a product is patented, that patent is public record, and at the end of the term of the patent others can use information in the patent to produce a competitive product. The development and fine-tuning of these standards can take a long time and in many cases proprietary standards are considered intellectual property of the corporation. The development of these standards is done at various levels of the corporation and access to them is sometimes limited to an as-needed basis.

6.6 Limited Consensus Standards

Limited consensus standards (sometimes referred to as "nonconsensus standards") are usually developed for the benefit of the members of a trade or technical association. Most, but not all, of these standards are accessible to the public. They are developed by various means such as an internal committee, a single individual, or a single company within the organization. The formal development process proceeds along the following lines:

- A request is made to the association by one of the members to develop a standard or a new rule within an existing standard for a desired entity.
- The staff forwards the request to an organizational governing body.
- The governing body locates a resource for developing the proposed standard or a rule within an existing standard. This resource can be inside or outside the organization.
- The written proposal is evaluated by the governing body. It may be sent out for a second opinion.
- The proposal is voted on by members of the association.

- The standard is developed by the selected resource.
- The standard is voted on by the members of the association.
- If approved, the new standard is published.

6.7   Standards Maintenance

Standards maintenance, or the process by which standards are kept current and usable, typically follows essentially the same process as that by which the standard was first developed. If a standard was developed internally by a company or trade association, then that process is usually followed for updates. For VCS to retain their favored status they must follow essentially the same process as that by which they were first developed. There are some differences in how organizations address the challenges of keeping a standard up to date (e.g., continuous updates, versus periodic updates), but for VCS the requirements for openness, transparency, due process, and consensus must still be met.

6.8   Summary

The variety of details in how standards are developed reflects the wide range of applications of those standards. When the stakes are particularly high, then the processes by which standards are produced and maintained tend to be highly rigorous to ensure a top quality product. For less critical applications the multiparty teams that develop the standard as well as the review processes used to ensure quality and public acceptability of the standards tend to be less formalized. If the process is not effective for its application, then it tends to evolve to one that is more successful. This is seen in those cases in which private standards, or even internal government technical standards have become VCS.

6.9   Case Study

A company develops software that uses artificial intelligence to help doctors diagnose a certain disease. The company has the following options:

1. Develop an internal standard to update the program.
2. Work with an existing consensus developing organization to develop a standard for updating this product that others can use.
3. Work with a private organization to develop a standard for updating this product where only the private members have access to updates.

Discuss the advantages and disadvantages of each of the above three options in the context of this product.

7

Types of Standards

7.1   Introduction

Standards are written differently depending on the culture of the organization developing them, the product or process they are defining, the capabilities of the organization implementing them, and other factors. There are numerous ways in which they can be categorized. Some of these were noted briefly in Chapter 1, and this chapter will address them in more detail. Some of these categorization approaches have been formalized in standards themselves, while some are used simply because they are helpful in understanding and approaching standards, whether for their use or their development. ISO/IEC Guide 2, Standardization and related activities – General Vocabulary, offers a number of useful ways of looking at standards, some of which are included in the following list.

7.2   Performance versus Prescriptive

7.2.1   Performance Standards

Performance standards are those that are written to define what must be accomplished. Writing a standard in this way allows the implementing organization a great deal of flexibility in how it chooses to comply. There are benefits to this approach, especially in those cases where the interfaces are minimal and what must be accomplished is easy to define. The risk is that not enough thought may be given to what is actually needed, and certain critical aspects of the product (process, etc.) may be left out of the standard. In this case, a fully compliant product may be produced and yet not fulfill what is truly needed.

Primer on Engineering Standards: Expanded Textbook Edition, First Edition.
Owen R. Greulich and Maan H. Jawad.
© 2018, The American Society of Mechanical Engineers (ASME), 2 Park Avenue, New York, NY, 10016, USA (www.asme.org). Published 2018 by John Wiley & Sons Ltd.

Cases in which a performance standard may be particularly desirable include the following:

- The technology is not well developed, or is evolving, such that it is too early to be more prescriptive.
- What must be accomplished can be readily defined in sufficient detail.
- A number of viable ways of accomplishing the task exist, and no one of them clearly dominates the market.
- It is possible to adequately characterize any interfaces, or there are none.

For example, a standard for developing automobiles could be produced as a performance standard. It could be written in broad general terms as follows:

> *Construct a car that will carry four people in safe, quiet, efficient comfort.*

Such requirements are all valid, but many more are needed. Carry them where and how far? How big are these people? How heavy? Do they have luggage? Is the car enclosed? Does it have seat belts? How safe is safe, and how do you achieve it? Quiet for the passengers, or for those in the surrounding environment? What does "efficient" mean? Is it made of a particular material? Does it have turn signals? The additional potential requirements go on and on.

Performance standards can be particularly useful when what must be accomplished is known, the ways of accomplishing the task have not been extensively examined and tested through experience, and there are not extensive interfaces or other restricting elements and requirements. They allow for flexibility and innovation. The challenge is often in determining and agreeing that their requirements have been met.

7.2.2   Prescriptive Standards

Prescriptive standards specify in detail what must be done. When the writer of the standard not only knows what must be accomplished, but also knows how it should be accomplished, then a prescriptive standard is likely to be appropriate. Most of the OSHA regulations are quite prescriptive. For example, in the section on overhead and gantry cranes they say things such as, "Fire extinguisher. Carbon tetrachloride extinguishers shall not be used," and "If sufficient headroom is available on cab-operated cranes, a footwalk shall be provided on the drive side along the entire length of the bridge of all cranes having the trolley running on the top of the girders." These are cases in which what is required can be clearly and unambiguously stated, and in which it makes sense to mandate consistency.

Aspects of a standard that might make a prescriptive standard desirable include the following:

- The field is well developed and it is not anticipated that the details specified would hamper innovation.
- What must be accomplished is understood and also how it is best accomplished.
- Processes have been well worked out and defined.

7.2.3   Component Standards

Component standards are the ultimate in "prescriptiveness." They are used when the exact product required is known. They specify all the materials, dimensions, and other essential characteristics of the part, such that a component that meets the standard will work in the application, and will be completely interchangeable with all other such parts. Component standards define pipe sizes, bearings, flanges and fittings, electrical components, and many other essentials of daily life. These are the standards that allow one to go to the store and purchase a light bulb, having full confidence that upon returning home again he/she will be able to screw it into essentially any light fixture in the house, turn on the switch, and have light. A person can buy faucets that will fit the sink, and locksets for the doors, with similar confidence.

This type of standard is used

- when interfaces and interchangeability are well understood and are critical,
- when mass production is applied by multiple manufacturers whose parts must fit together.

If innovation is to occur in a situation such as this, it will likely either come in the form of incremental change or result in a new standard for other configurations that will fit together with each other. Thus a different light bulb interface would have a new standard, or section in an existing standard, to define the critical characteristics of that interface.

7.2.4   Hybrid Standards

Other than for component standards, it is very common to use a combination of performance and prescriptive requirements. This is practical when some aspects of a product or process are, or can be, very well defined, while others remain open.

Returning to the example of the automobile above, certain aspects are "non-negotiables," while others are left entirely to the designer. Antilock brakes are required, but the details of the design are not fully defined by regulation. Many specific components are purchased to component standards, others are designed to prescriptive standards, and some aspects of the design are left almost entirely up to the engineer. This hybrid approach tries to take advantage of the strengths of each type of requirement, while minimizing the risks and drawbacks.

Similarly, the AIAA standard for composite overwrapped pressure vessels is very specific about some things such as safety factors and acceptance requirements, but it allows flexibility in how stress analyses are performed, something that in the ASME BPVC is fairly tightly defined in most cases.

Hybrid standards are often preferred because their use permits a great deal of specificity when needed while retaining the advantages of performance standards for other aspects of a product.

7.3 Geographical, Political, or Economic Extent

It is sometimes useful to categorize standards on the basis of the extent of their influence. ISO/IEC Guide 2, Standardization and related activities – General Vocabulary, offers the following categories to describe the extent of influence of various standards.

International Standardization: Standardization in which involvement is open to relevant bodies from all countries.

Regional Standardization: Standardization in which involvement is open to relevant bodies from countries from only one geographical, political, or economic area of the world.

National Standardization: Standardization that takes place at the level of one specific country.

Note: Within a country or a territorial division of a country, standardization may also take place on a branch or sectoral basis (e.g., ministries), at local levels, at association and company levels in industry, and in individual factories, workshops, and offices.

Provincial Standardization: Standardization that takes place at the level of a territorial division of a country.

Note: Within a country or a territorial division of a country, standardization may also take place on a branch or sectoral basis (e.g., ministries), at local levels, at association and company levels in industry, and in individual factories, workshops, and offices.

7.4 Mandatory or Voluntary

As previously noted, compliance with some standards is mandatory while that with others is voluntary. This book uses the term "jurisdictional standard" to refer to those standards that are mandatory and "voluntary consensus standard (VCS)" for those that are voluntary. Again, in some cases jurisdictional standards will make use of incorporation by reference to require compliance with certain otherwise voluntary standards. This practice is addressed in more detail in Chapter 5.

7.5 Consensus versus Nonconsensus

Interest groups and large companies often develop their own standards. While these must conform to the well-established national and international standards if compliance with them is required or claimed, certain process details necessary for use or production often remain proprietary. These details often deal with procedures and guidelines for design, manufacturing, testing, documenting, and quality control. As such, they are part of the closely guarded intellectual property (IP) that keeps a company competitive.

These standards typically do not meet the ANSI requirements for VCSs (one exception is UL, which does publish VCS). The distinctions among these categories are discussed in more detail in Chapters 3–5.

7.6 Purpose

There are many categories that describe various purposes of standards. Among these are fitness for purpose, quality, compatibility, interchangeability, and safety. A number of other categories could be identified as well. Classifying standards in this way can be helpful, but often a single standard will address more than one of these issues. Still, this classification approach can be particularly useful in deciding where to draw the lines between two or more standards.

As an example, ASTM A-269, Standard Specification for Seamless and Welded Austenitic Stainless Steel Tubing for General Service, includes only a small part of the total requirements for this product. It references five other ASTM standards covering such topics as practices for detecting susceptibility to intergranular attack, requirements for plate, sheet, and strip that are used to make welded tubes, and general requirements for tubes. By breaking down the requirements in this way, a great deal of redundancy is avoided, since the referenced standards are also referenced by many other higher level standards.

7.7 Subject

Looking at standards from a subject perspective, such as testing, product, process, service, interface, and data to be provided, gives another view. This approach can be helpful in making certain that a standard has addressed everything it should, whether internally or by reference. It can also simplify the process of identifying applicable standards for various parts of an operation.

7.8 Surprise Consequences of a Successful Standard

The importance of standards can be illustrated by the case of IBM's entry into and exit from the personal computer market. IBM, a large computer company up to that point mostly interested in the main frame computer market, decided to join the personal computer market in 1981 by introducing its own computer. It also decided to use off-the-shelf electronic components with a modular design. The breakthrough in the personal computers came when this large computer company threw its weight behind the development of the back plane bus architecture standard that independent vendors could use to interface their product to the CPU. These products included the RAM, video card, sound card, hard drives, modem and Ethernet cards for communications, A/D and D/A converters for industrial control, and other internal devices.

Externally, user interface devices such as keyboard, mouse, touch pad and joystick, and other peripherals such as printers would interface through the standard serial and parallel ports. The only part that IBM made proprietary was the BIOS (basic input/output system). This was later developed by other vendors and made compatible with the IBM design, making it possible for other PC compatible (clones) to be manufactured by other companies, becoming the de facto, and ultimately the actual, standard. So successful and popular was this open bus architecture that when IBM introduced the more advanced but proprietary Micro Channel Architecture (MCA) in its personal system computers, it failed to secure sales and vendors refused to cooperate. In 2005, IBM sold its personal computer business to another company and exited the market.

An example of development of a component at first in the absence of standards, then with standards catching up is the Universal Serial Bus (USB) used in computers to communicate with various devices. Initially, computer manufacturers used only ports that were compatible with their own peripheral equipment in an effort to control the market. However, the explosion in the peripheral market, reinforced by consumer demand, forced the computer manufacturers to consider adapting ports to fit peripherals made by other manufacturers. Intel came up with a protocol framework for USB that allows various devices to send whatever data they desire. Acceptance of the USB was slow in the beginning when manufacturers were skeptical that they needed to accept it. When a USB was plugged in a computer a message would flash indicating "a driver is needed in order for this device to work." However, as time went on, all of the operating systems (Windows, Mac, Linux, etc.) incorporated the needed drivers.

Initially, a small USB group was formed to develop a standard. Many companies joined the group later, once the benefits of USB became obvious. Today, it is rare to plug in a device and NOT have a driver available to transmit data. The data consists of a well-defined header packet that is standardized by a USB group. The group also standardized the physical dimensions, then the mechanical and electrical, and then the protocol level. The standard was revised many times over the years to take advantage of hardware and software changes that allowed for significantly greater data transmission speeds and to fend off competition from other interface protocols that were also evolving, with issues 1.0, 1.1, 2.0, 3.0, 3.1, and the current USB3.2. The current standard for the USB is over 600 pages long (www.usb.org).

The following is a partial list of topics of standards that have since been developed in the relatively new fields of computer hardware and software:

Hardware	
ACPI	Advanced Configuration and Power Interface
AGP	Accelerated Graphics Port
ATX	Advanced Technology eXtended
DVI	Digital Visual Interface
EISA	Extended Industry Standard Architecture
Fire Wire	(IEEE 1394)
HDMI	High Definition Multimedia Interface
IDE	Integrated Drive Electronics. Extended to ATA/ATAPI
ISA	Industry Standard Architecture
PCI	Peripheral Component Interconnect
SATA	Serial ATA
SCSI	Small Computer System Interface
UDI	Unified Display Interface
USB	Universal Serial Bus
VESA	Video Electronics Standards Association

Software	
API	Applications Programming Interface
CGI	Common Gateway Interface
HTML	Hyper Text Markup Language
HTTP	Hypertext Transfer Protocol
MPI	Message Passing Interface
ODF	Open Document Format
PDF	Portable Document Format
PNG	Portable Network Graphics
POSIX	Portable Operating System Interface (IEEE)
PS	Post Script
RTF	Rich Text Format
SQL	Structured Query Language
SVG	Scalable Vector Graphics
TWAIN	Technology Without An Interesting Name
UDF	Universal Disk Format
Unicode	Universal Code for text transfer
WAP	Wireless Application Protocol
WML	Wireless Markup Language
XML	Extensible Markup Language

7.9   Summary

A number of approaches are used to categorize standards. Thinking about standards in this way can be helpful in determining how to structure a standard or group of standards, as well as ensuring their completeness. While it is not a topic of everyday discussion, understanding the various types of standards as looked at from different perspectives can also help in their effective use.

7.10   Case Study

Describe a standard that might have helped ease the transition to common use of the USB, discussed in Section 7.8, and identify performance and prescriptive aspects of such a standard.

8

Conformity Assessment

8.1 Introduction

While standards are essential to making widespread consistency of products and processes possible, the confidence in actual compliance with the standards is made possible by conformity assessment. The importance of conformity assessment is reflected in the fact that it is almost always included in any policy statements regarding standards.

There are many definitions of conformity assessment depending on the field of application. One definition, as it pertains to engineering applications, is given by ISO/IEC Guide 2:

> *Any activity to determine, directly or indirectly, that a process, product, or service meets relevant technical standards and fulfills relevant requirements.*

A second definition, similar but somewhat more explicit, is provided by OMB Circular A-119, the direction from the Office of Management and Budget, Office of the President of United States, as implementation of the National Technology Transfer and Advancement Act of 1995:

What is Conformity Assessment?

"Conformity assessment" is a demonstration, whether directly or indirectly, that specified requirements relating to a product, process, system, person, or body are fulfilled. Conformity assessment includes sampling and testing, inspection, supplier's declaration of conformity, certification, and management system assessment and registration. Conformity assessment also includes accreditation of the competence of those activities.

Primer on Engineering Standards: Expanded Textbook Edition, First Edition.
Owen R. Greulich and Maan H. Jawad.
© 2018, The American Society of Mechanical Engineers (ASME), 2 Park Avenue, New York, NY, 10016, USA (www.asme.org). Published 2018 by John Wiley & Sons Ltd.

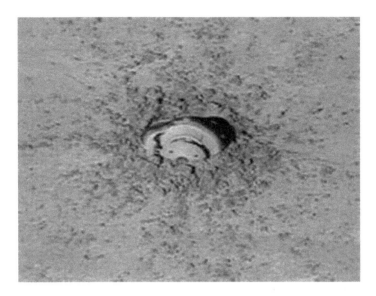

Figure 8.1   Genesis space probe sample return capsule
(Courtesy of NASA)

Conformity assessment programs are vital for acceptance and use of engineered products because they ensure not only compliance with the standards and requirements used during manufacturing and/or operation (verification), but also that the correct requirements were put in place in the first place (validation). Failure to use conformity assessment programs effectively may lead to catastrophic results such as the one shown in Figure 8.1: Genesis was a space probe intended to sample the solar wind. While it was successful in doing so, the return to earth of the samples was marred by failure of parachute deployment, caused by a G-switch that was installed backwards. The chair of the NASA mishap investigation board stated that a pretest procedure skipped by the supplier could have easily detected the problem.

8.2   Users of Conformity Assessment

Conformity assessment in engineering is usually applied by one of the following interested parties:

1. Producers and manufacturers.
2. Users and their designated agents
3. Inspection/insurance agencies and inspection jurisdictions

4. Governmental agencies such as the United States Nuclear Regulatory Commission (NRC) and the Environmental Protection Agency (EPA) that may have an interest in the safety of certain engineered products and their operation.

8.2.1   Producers and Manufacturers

Producers and manufacturers have multiple reasons to be interested in the conformity of their products. A product that does not meet the requirements is not going to be welcomed by customers whether it simply fails to do what the customer wants or fails to comply with the required standards and codes. Producers and manufacturers demonstrate their compliance with requirements by following conformity assessment programs. An example is the assessment procedure, referred to as the Quality Control Procedure, of manufacturers showing their compliance with the requirements of the ASME Boiler and Pressure Vessel codes. ASME publication CA-1 titled "Conformity Assessment Requirements" gives detailed instructions for the quality control procedures of manufacturers. Similar conformity procedures are required for many manufactured products ranging from cranes to computer chips to threads and pumps.

8.2.2   Users and their Designated Agents

Many users must have in-house conformity assessment programs in order to ensure that they operate engineered equipment safely, consistently, and reliably. An example is hospitals where complicated equipment such as CAT scanners, sterilizers, and X-ray machines are used on a daily basis for functions in which the patient's welfare depends on proper operation. Hospitals usually have detailed conformity assessment programs to assure that personnel operating specific equipment are qualified to do so and that proper procedures are followed. They also make use of outside accrediting organizations to perform independent conformity assessments.

8.2.3   Inspection/Insurance Agencies and Inspection
Jurisdictions

Companies that insure products such as boilers, pressure vessels, and certain manufacturing and processing facilities protect their interests and those of their clients by performing conformity assessments. They typically have authorized inspectors or other assessors who need to be familiar with the details of operation and maintenance of the equipment in order to perform their conformity assessments effectively. The inspectors must comply with a number of requirements, including certification, continuing education (periodic classes to ensure currency of their knowledge), and minimum experience levels. Accordingly, insurance companies have detailed conformance assessment programs to monitor and ensure compliance with internal and external requirements of authorized inspectors performing assessments for them.

Some cities, counties, and states have their own authorized inspectors. In such situations, these jurisdictions must likewise have conformance assessment programs in place.

8.2.4 Governmental Agencies

Conformity assessment programs may also be required at such governmental agencies as the NRC for nuclear components, the EPA for commercial equipment, and the FAA for critical flight programs. Highly specialized personnel are needed for these jobs and a conformity assessment program also ensures proper training and performance of such personnel. US Government regulation 15CFR287 has guidance in this area. It also includes the "legal" definition of conformity assessment (and a number of other related terms) as it applies to the US Government. While written as guidance, and fairly generally, it delineates conformity assessment responsibilities of federal agencies.

8.3 Applicability of Conformity Assessment

Conformity assessment may pertain to the following:

1. Employees
2. Processes
3. Products
4. Services
5. Systems.

8.3.1 Employees

Conformity assessments are applicable to designated employees of manufacturers who manage the in-house conformity assessment programs, as well as to many of those performing critical operations. Again, using the case of pressure vessel manufacturing, an employee assigned to overseeing the conformity assessment program for the company must meet certain requirements. For example, in addition to possessing certain capabilities, this employee, who works in the plant, must not report to the plant manager but to another arm of the company since her/his responsibilities include monitoring both the plant and the plant manager for conformance. If this employee were to report to the plant manager, there would be an obvious conflict of interest that could make it impossible for the employee to perform her/his role successfully.

8.3.2   Processes

Failure to follow an established manufacturing or operating process may lead to undesirable consequences. This happens, for example, in welded products when the welder does not follow the exact welding parameters related to such items as voltage, current, and speed of welding. It may not be possible to nondestructively determine certain critical properties (low temperature toughness, crack growth rates, etc.) of the weld so the quality of the final product is dependent on effective control of the process. For this reason, conformity assessment programs must always be in place to assure that the process is followed. Similarly, the process of integrating the design, acquiring raw materials, and assembling the product must follow certain procedures to avoid unwanted results.

8.3.3   Products

There are numerous instances in which a product has been shipped to a given location only to have the purchaser discover that it is unacceptable, whether failing to comply with user needs or violating existing local law, standard, or other requirement. The cost of remedying such a situation is sometimes very high. Thus, it behooves the organization specifying the product to have in place a conformance program that includes either checking local requirements regarding acceptance of such products or a procedure for contacting appropriate parties for product approval prior to shipping, along with the program that ensures the quality of the product itself.

8.3.4   Services

Many products such as boilers, bridges, and buildings must be serviced in place and cannot be shipped back to the fabricator for repairs. Servicing such products requires recognition of the local standards and requirements. Personnel, equipment, and procedures must be properly qualified in accordance with the standards applicable on site. To accomplish this, a conformity assessment program needs to be in place to avoid unnecessary delays and cost to the project.

8.3.5   Systems

Conformity assessment is vital to the operating system of a company for product manufacturing and/or operation. For example, the sequence of startup and shutdown of many pressure vessels in a refinery or chemical plant is extremely important in maintaining the integrity of the equipment. A somewhat different example of the need for a conformity assessment program for operations within a company relates to repairing equipment whose operation could be hazardous to maintenance personnel. To perform the repair safely, all sources of energy must be cut off, using a procedure referred to as "Lockout–Tagout." Sources of electrical power need to be shut off and valves of incoming and outgoing pipes connected to the equipment being repaired need to be closed, and the equipment may need to be properly ventilated to allow personnel to enter or otherwise work on the unit safely. Some of these power sources and valves may be a long distance away from the unit and operators may not be aware of their location without a detailed plan, yet their lives may depend on the switches and valves being in the proper configuration. An appropriate conformity assessment program is used to ensure that they are.

8.4   Verification and Validation Process

One manifestation of the conformity assessment process is referred to as Verification and Validation (V&V). This terminology is frequently used in software development.

One definition of V&V, given by IEEE standard 610, is

> *Validation: The process of evaluating software during or at the end of the development process to determine whether it satisfies specified requirements.*
>
> *Verification: The process of evaluating software to determine whether the product of a given development phase satisfies the conditions imposed at the start of that phase.*

Thus, validation deals with the question, "Have we correctly specified the product we are planning to build," while verification deals with, "Does the product we are building meet the specification?"

Simply put, validation consists of processes implemented throughout the life cycle to ensure that the product being produced is correctly defined and specified, while verification consists of processes, again throughout the life cycle, to make certain that the product is correctly built to meet the specification, requirements document, statement of work, etc.

Initially, the V&V process was used to assure that complex software programs do what they are supposed to in order to produce correct and accurate results. In recent years, this process has been formalized as a quality assurance process used in software development to ensure that the final product does what it is intended to. While the development of appropriate requirements for a product (software or otherwise) may be for the most part manageable, the complexity of software products, often with millions of lines of code, has become so great that it is often not practical, perhaps not even feasible, to check every possibility of how the product might work. V&V is one tool used to greatly enhance the chances that the product will work as planned.

As new industries develop and grow, it is common for standards to ensure the quality and/or safety of the products of those industries to develop and grow with them. An example of this has occurred with the software industry. ISO/IEC 12207:2008, Systems and Software Engineering – Software Life Cycle Processes, is an international standard used to ensure the quality of software products. It is notable that this standard was first issued in 1995 (relatively recently) and that it first came out as an international standard (as opposed to proprietary, or an individual country standard). The current standard includes 43 system and software processes dedicated to providing a common structure so that stakeholders in the software development process can use a common language in product development, procurement, and assessment, established in terms of the processes.

A portion of ISO/IEC 12207 deals with V&V, two of the processes used to ensure that software products are correctly specified and that they are built to meet the specification. These processes are a formalization by the software industry, tailored to the needs and capabilities of that industry, of an approach that has long been used in a wide range of industries to ensure that a product being developed meets its intended purpose. Whether in the software industry or elsewhere, when a complex product is produced, it is necessary to plan processes to ensure quality and to implement them throughout the life cycle of the product.

ASME is another organization heavily involved with writing standards related to V&V. Standards such as V&V in Computational Solid Mechanics, V&V in Computational Simulation of Nuclear System Thermal Fluid Behavior, V&V in Computational Fluid Dynamics and Heat Transfer, and V&V in Computational Modeling of Medical Devices are examples of the standards being developed in this field by ASME.

Currently, the V&V planning process is used for software development as well as other product development. The process must assure that key parties involved with the product development discuss, at key points during the development state, such items as progress, path forward, and adjustments to the initial plan based on progress of the product. The nature of the discussion is negotiations and consensus among the primary stakeholders to assure a satisfactory product.

8.5   Conformity Assessment Organizations

There are numerous organizations that offer third-party conformity assessment reviews. Some of these organizations are listed here:

Acronym or Abbreviation	Name and web address	Major fields of action
ACB	American Certification Body (acbcert.com)	Certification services for wireless equipment
ASCI	Automation Standards Compliance Institute (isa.org)	Charter includes compliance assistance for software and hardware products
ASME	American Society of Mechanical Engineers (asme.org)	Approves manufacturers for fabricating boilers, pressure vessels, and nuclear components. Also publishes ASME CA-1-2014 for Conformity Assessment Requirements
ASSE	American Society of Sanitary Engineering (asse-plumbing.org)	Plumbing product certification
Bureau Veritas	Bureau Veritas Group (bureauveritas.com)	A global company in testing, inspecting, and certifying a wide range of consumer products
CSA	CSA Group (csagroup.org)	Conformity services in the consumer products area
Dekra	Dekra Certification, Inc. (dekra.com)	Vehicles, wind turbines, and aviation products
DQS	DQS, Inc. (dqsus.com)	Certifies management systems
FM Approvals	FM Global Corporation (fmapprovals.com)	Fire prevention, sprinkler systems, and insulation products
ICAP	IEEE Conformity Assessment Program (ieee.org)	Electrical engineering, electronics, and telecommunications

Acronym or Abbreviation	Name and web address	Major fields of action
ICC-ES	ICC Evaluation Service (icc-es.org)	Performs conformity assessment and evaluation of building products in the construction industry
INTERTEK	Intertek, Inc. (Intertek.com)	Conformity assessment as well as testing and certification in various industries such as chemical, construction, energy, and transportation
NBBI	National Board of Boiler and Pressure Vessel Inspectors (nationalboard.org)	Approves manufacturers for fabricating and repairing boilers, pressure vessels, and nuclear components
UL	Underwriters Laboratories (ul.com)	Various areas such as electrical, fire, and HVAC

8.6 Summary

While conformity assessment has only been named and formalized as such in recent decades, the field has been in place for much longer. Any organization using a formal or informal program to ensure the quality and compliance of its products or processes is employing elements of conformity assessment. Many companies and government agencies have long had formal and well-documented programs to ensure that what they do or produce is properly specified and that the actual work and product comply with that specification. The formalization of the field of conformity assessment has resulted in higher expectations on the part of consumers that the product they purchase will really deliver what they want.

8.7 Case Studies

Case Study 1

A manufacturer wishes to produce engineered hospital beds.

What are the key items it must consider when developing the conformity assessment program regarding (1) product features, (2) manufacturing details, (3) documentation?

Case Study 2

A company has determined through business studies, surveys, and focus groups that customers for the products that it sells would be willing in some cases to pay premium prices for delivery within 2 h of ordering. It wishes to produce the drone version of the grocery store delivery boy of the 1950s.

Discuss the components of a drone delivery system and the conformity assessments that might apply to each.

9

Standards Interpretation and Relief

9.1 General Discussion

As described in Chapter 3, standards are generally produced by experts in the field using a deliberate and rigorous process. The writers and the developing organizations make every effort to produce a product that is clear, concise, and complete. Unfortunately, knowledge is incomplete and even with the best of intentions it can happen that there is a need for further interpretation, explanation, or relief. Both the interpretation process and the process by which relief is acquired depend on the type of standard, the way it was adopted, and whether the standard itself includes provisions for relief.

"We're from the Galactic Bureau of Standards, and this planet is *way* out of compliance!"

(Courtesy of CartoonStock.com, www.cartoonstock.com)

Primer on Engineering Standards: Expanded Textbook Edition, First Edition.
Owen R. Greulich and Maan H. Jawad.
© 2018, The American Society of Mechanical Engineers (ASME), 2 Park Avenue, New York, NY, 10016, USA (www.asme.org). Published 2018 by John Wiley & Sons Ltd.

There are times when an organization or a jurisdiction may need to take an exception to a recognized standard for product manufacturing or operation. This is usually formalized and documented with a variance or waiver (sometimes also referred to as deviations or tailoring). Waivers often involve complex issues and should only be pursued after careful evaluation of consequences as well as the amount of effort involved.

Use of the terms "variance" and "waiver" is somewhat ambiguous. Variance is generally fairly consistently defined, but in common usage the term "waiver" is frequently used in its place. The terms will be used interchangeably in this book:

Tailoring: The fitting of a standard to the particular needs, values, or priorities of a jurisdiction, organization, or program. Details of tailoring are given in Section 9.5.

Variance: An exception to a standard, regulation, or ordinance, authorized by the appropriate authority having jurisdiction, such as the Occupational Safety and Health Administration (OSHA), a planning commission, zoning board, county commissioners, or city council if jurisdictional. In some cases, particularly for nonmandatory standards such as voluntary consensus standards (VCS), variances may simply not be issued, and in others a deviation from the standard is addressed through what might be considered an interim or temporary change to the standard such as an ASME "Code Case." Moreover, while the term "waiver" is often used for exceptions to standards, "variance" is typically the more accurate expression. The word "waiver" is often interpreted to mean that the whole requirement can be ignored, while the usual use case is that something varies from the requirement but the issue is still addressed. Additional analysis may have been performed as justification for acceptance of a lower safety factor, or other measures taken to ensure an equivalent level of safety.

For the purposes of this book, mandatory standards are considered to be only those standards with force of law themselves (the law, regulation, ordinance, etc.). VCS invoked by the law, regulation, or ordinance are addressed as nonmandatory because not every jurisdiction requires compliance with them. In the current discussion, corporate policies and standards are considered nonmandatory, since the corporation can choose to accept noncompliances from them, or can change them, at any time.

9.2   Standards Adoption

Standards cover a wide range of activities, are produced by many different organizations, and are adopted in a number of ways. Company internal standards can be made mandatory within the company either through a formalized process or by simple dictate from a person with suitable authority. Jurisdictional standards become mandatory through action of legislative bodies, have the effect of law, and may in addition invoke application of VCS. VCS become the standard for companies in several different ways, including company internal decisions, prerequisites for purchase of insurance, references in jurisdictional standards (see the Section 5.3), and through direct legislative action. Those VCS that are imposed by regulations usually come to have essentially the same authority as the implementing regulation, and are usually accepted in their entirety, often without further comment. Those imposed by direct legislative action or adoption are sometimes accepted with modifications, deletions, or additions by the legislative body (see Section 9.5.2).

9.3   Effect of Noncompliance with Standards

A variety of factors determine the effect of noncompliance with a standard. These include the type of standard, whether it is truly accepted as an industry standard, whether an organization has adopted it as policy, whether it is mandatory, whether it is regulatory, and how the noncompliance came about.

Under the OSHA regulations, an employer is responsible for the safety of its employees. That entity (assume for discussion that it is the system owner) therefore has final responsibility with respect to OSHA, but it is usual for the owner to contract with various other organizations for the design and construction of a facility. While final safety responsibility remains with the owner, the form of the contract by which the owner purchases design and construction services can affect how responsibility is allocated among the parties performing design and construction.

Unless the owner specifies all details of what standards are to be used in design, it is normally the responsibility of the designer or the organization in charge of the design to comply with jurisdictional requirements where the product is installed or operated. These requirements may specify something other than the latest edition of a code or standard. Also, many standards reference other standards within their text. These internally referenced standards also may not be the latest revision. Conflicts may exist between two national standards regarding details of construction and the designer needs to be aware of which one controls in the jurisdiction where the product will be used. This situation sometimes occurs in the case of alterations to pressure vessels where the original code of construction, say the ASME code, specifies certain radiographic examination, hydrotesting, and other NDT requirements, while the jurisdiction using such standards as the NBIC or API 579 codes requires other provisions due to the particular difficulty or nature of the alteration.

Another example of a possible conflict is the use of building codes in construction. Until 1994 there were at least four major building codes used in the United States. The BOCA code (Building Officials Code Administrators) was used mainly on the East coast and in parts of the Midwest, the SBCCI code

(Southern Building Code Congress International) was used mainly in the South and Southeast, and the UBC (Uniform Building Code) and ICBO codes (International Conference of Building Officials) were used mainly on the West Coast and in parts of the Midwest. The designer had to be aware of all of these codes and their regions of applicability. By 1997, all of these codes were combined into one code called the IBC (International Building Code). However, some places still have unique requirements. For example, the State of California has its own building code called the CBC (California Building Code) that must be complied with, addressing special requirements for earthquakes and for green construction.

Situations often arise where equipment and components such as pressure vessels built to an international standard, such as the ISO, are imported into the United States. In such cases, the laws in the given State are not complied with since the majority of the states in the United States have laws specifying the ASME code as the controlling standard for pressure vessels. The laws typically require that such equipment not be operated until a "variance" has been approved. Some states require this variance to be obtained from the State prior to fabrication of the equipment. Typically, the variance is granted by the Board on Pressure Vessels for the State and a "State Special" item number is issued by the Chief Boiler Inspector of the State to be stamped on the name plate of the equipment. In the deliberation regarding whether to issue the variance, the governing body considers the following factors:

- Does the foreign standard cover all aspects of manufacturing such as design, materials, inspection, testing, and quality control?
- Is the foreign standard internationally recognized?
- Is the foreign manufacturer qualified to build in accordance with the foreign standard?
- Are the differences in requirements between the foreign code and ASME detrimental to safe operation of the equipment within the State?
- Do additional technical requirements need to be applied to bring the product to an acceptable level of safety?

The above process can be lengthy and the user should include a significant amount of time in the schedule for obtaining the required variance.

9.3.1   New Products

Aside from the technical and cost issues involved, an important component in the decision whether or not to comply with a standard is the motivation for applying it in the first place. Sometimes, a nonconforming product or process is simply unsuitable and no relief from a standard should be pursued. In some cases, a deviation from a standard may have essentially no consequences, with no one outside the organization ever becoming aware of, or caring about it. Other times, while performing its function in a suitable manner, a product will look, feel, or operate differently from intended and the company itself rejects the waiver because of the image they wish to project to their customers. In still other cases, the standard from which the deviation is proposed is jurisdictional. If this is the case, then the company must follow whatever formal relief process is required by the jurisdictional agency. If it does not, it will be at risk for any legal sanctions that apply (usually fines, in some cases possibly criminal charges).

The related issue of a decision by a jurisdiction not to require compliance with a generally accepted standard illustrates some other consequences of noncompliance with a respected and time-tested VCS. The State of Texas has decided not to mandate conformance of pressure vessels with the ASME Boiler and Pressure Vessel code (BPVC). This is not a waiver or variance, but it involves many of the same issues as a waiver. Companies operating in Texas may install and operate pressure vessels that are not code stamped without violating state law, but once this has been done issues other than the lack of a requirement for code stamping become primary.

Liability can play a major role in decisions regarding equipment purchases. In case of a failure, experts will likely compare the design and operation of the failed equipment to available products meeting a recognized standard. Deviation from the standard will almost certainly be questioned as to its appropriateness with respect to safety and engineering judgment. Injured parties may sue for damages, and a jury is likely to be sympathetic to a plaintiff who can demonstrate that the responsible party was using equipment that did not meet generally accepted, though perhaps not mandatory, industry standards. While this book points out a few areas that may involve legal problems, it is not intended to provide legal advice. If there is a concern in this area, it is important that the user of the standard or product seek out appropriate legal assistance.

In cases in which the OSHA regulations do not specify the industry standard, OSHA can still cite an employer for not maintaining a safe workplace if there is an incident and it can demonstrate that the employer was not meeting industry standards for its operation. Also, while Texas does not require ASME code pressure vessels, in its 2006 report on the Marcus Oil and Chemical Tank Explosion, the United States Chemical Safety Board, which investigates such incidents, recommended that the City of Houston amend city building ordinances to require that newly installed pressure vessels comply with the ASME BPVC, Section VIII, and that repairs comply with the National Board Inspection Code.

In addition to the concerns above, equipment in many plants must be insured prior to operation and most insurance companies require compliance with VCS prior to insuring the equipment (Some insurance companies take an active role in the development and maintenance of codes and standards, and even in the inspection of boilers themselves, both in plant for operation, and during manufacture.). Thus in most cases pressure vessels operated by responsible organizations even in Texas wind up conforming to the ASME code although the state does not require it.

Finally, it should be noted that there are certain types of standards from which waivers are probably rarely or never used. Fasteners, for example, either meet or do not meet their standards. Gears either meet or do not meet. Many noncritical applications use fasteners, for example, that are probably nominally in compliance with a standard. In many cases there is probably no verification of such compliance, and because often loads are not high and failure is not critical, many noncompliant fasteners may be in use. No waiver would be processed for fasteners in such an application, and conversely, a substandard component would simply be rejected if identified in a space flight or other critical application.

9.3.2   Post-Manufacturing Noncompliance

9.3.2.1   General Discussion

Products manufactured in accordance with a standard may become noncompliant with the standard through aging, wear and tear, damage, or a change in the standard by the standards developing organization. Such noncompliance should be evaluated by the owning/operating organization, and an assessment should be made of the consequences of the noncompliance, followed by a decision as to whether or not to upgrade the product in order to comply.

9.3.2.2   Product Degradation

The consequences of degradation with time and usage may vary, depending on the type of system and its application. Two examples will be briefly discussed here: (1) wear in the lead screw of a milling machine, and (2) corrosion of the wall of a piping system.

Lead screws in a milling machine are those screws that drive the table on a milling machine left and right, in and out, or up and down. Such a component is manufactured in accordance with an appropriate thread standard, but it is also subject to wear. Typically, this wear takes place mostly in the center portion of the screw (because a mill is typically used for machining small components mounted at the center of the table). If this occurs, the average pitch of the screw over its length will not change, and depending on the type of wear, even the actual pitch may not change. There may be a change in the effective pitch of the screw, however, since the table location is driven by one side of the screw thread in one direction, and the other side in the other direction, and some portions of each side of the screw will be affected more than others.

If the sole means of ensuring accuracy of table position is by keeping track of turns of the screw (as in conventional machining and low-cost numerical control conversions), then it is likely that accurate positioning of the table will suffer.

Depending on the size of the component being produced and various choices made by the machinist, component errors may be minimal, but in certain cases the amount of wear on the screw can result in one-for-one variance of the work piece from its intended configuration (If the component is small enough that it is machined entirely within a portion of the screw with equal wear, tolerance errors due to lead screw wear may be insignificant or nonexistent.). Of course, if location is determined by a separate linear encoder, then the machine can compensate for the wear and the work piece will not be significantly affected by this type of wear. In any case, until the wear is so significant that a problem arises involving out of tolerance parts, such wear may not be a concern. In a case like this, a company can make a rational decision to continue using a lead screw long after it is out of tolerance without in any way affecting its product. In addition, even if the product is affected, the company may choose to do nothing because the effects are not catastrophic. Effects on the quality of the product may be minor, if they exist at all, and if decisions on use of such equipment are made judiciously, it is likely that no laws are violated and no one is harmed.

In the case of a corroded pipe wall in a pressurized piping system, corrosion of the wall removes material that may be needed to ensure system integrity. Piping systems are constructed in accordance with appropriate standards depending on their service. Further, those used in a corrosive environment are typically either built of corrosion-resistant materials, or they are built with a corrosion allowance that is intended to provide a reasonable life for the system before safety factors are reduced to the extent that the pipe must be removed from service. The consequences of a piping failure due to degradation are more likely to involve significant injuries, property, and possibly environmental damage than those from something such as wear of a lead screw in a milling machine (the same could be said of wear in a mobile crane). Similarly, the risk associated with such a failure is commensurately greater, and the consequences of a piping failure due to reduced wall thickness are further multiplied by potential legal consequences due to failure to comply with regulations mandating compliance with VCS.

Thus, in each case a company has a need to monitor the state of its equipment as it is used, but in one case a component becoming noncompliant will entail much greater risk than in the other.

9.3.2.3   Noncompliance Due to Changes in Standards

It is not unusual for a component or an installation to become noncompliant after purchase or installation due to changes in a standard. A common situation in which this occurs is in electrical wiring. The National Electrical Code (NEC) is continually being updated to keep pace with technology and increased demands for the safety of society.

In most cases, when a new edition of a standard is issued, a jurisdiction does not require update of existing installations to meet the requirements of the new edition. Thus when the NEC added a requirement for arc-fault breakers for residential electrical circuits serving bedrooms, while millions of homes became immediately noncompliant with the current edition of the standard, most were not noncompliant with laws, and nearly no one had to retrofit their circuit breakers. First, there is almost always a lag between issuance of a code or standard and its implementation as mandatory by a jurisdiction, and second, even when a new edition of a standard is made mandatory, it is highly unusual that a jurisdiction requires upgrade of existing facilities or equipment unless a major modification or upgrade to a building or system is made.

Another perspective on this issue is illustrated by a particular change to the ASME BPVC. With the addition of Figure UCS-66 and associated textual changes in 1989, the ASME BPVC made significant changes in material toughness requirements and with respect to which service would require testing. When this occurred, while jurisdictions did not specifically require re-evaluation or upgrade of systems, there was ample reason for pressure vessel users at least to assess their pressure vessels. In this case, the change reflected a new awareness of a hitherto unrecognized hazard (brittle fracture at temperatures higher than previously realized), and a responsible user aware of the change would want to ensure the safety of its systems and personnel, and would perform at least a cursory evaluation in order to make rational decisions regarding risk.

9.3.2.4   Post-Manufacturing Compliance

Most engineering standards are created as requirements or guidance for manufacturing of new equipment, but equipment subject to service-related degradation may be evaluated against them as well. Thus either the milling machine lead screw or the piping discussed above will be, if necessary, compared to the original standard to which it was produced. In the case of the machine lead screw, there is no jurisdictional requirement to meet the standard, and so a company is free to evaluate any nonconformance, determine its consequences on the components produced by the machine, and decide whether to repair or replace the screw. Certain piping systems are subject to jurisdictional requirements that impose codes and standards, and compliance with the standard in those cases is no longer a matter of choice.

Pressure vessels and piping are somewhat unique, in that there are post-construction codes developed by the National Board of Boiler and Pressure Vessel Inspectors and the American Petroleum Industry that apply and that have historically been distinct from the ASME codes, which applied only to new construction. These standards provide means of evaluating pressure vessels and piping subject to corrosion, erosion, fatigue, and other service-related degradation for which restoration to like-new condition may not be possible, and they provide guidance as to whether the equipment must be repaired or replaced, or can be used as is. It is only recently that API and ASME have begun to come together on addressing post-construction issues. This is reflected in API 579-1/ASME FFS-1 Fitness for Service.

Repairing and altering such equipment currently may require multiple waivers and variances from the existing standards. Various jurisdictions in the United States and Canada have provisions for repairing and altering pressurized equipment constructed to an earlier edition of the ASME BPVC. Numerous methodologies currently exist for repairing and altering older pressurized equipment. If one of these is followed, then insurance companies as well as jurisdictions, whether local, state, or federal (as, e.g., OSHA), will generally be satisfied.

9.4   Standards Interpretation

9.4.1   Informal Processes

While some Standards Developing Organizations (SDO) (e.g., ASME; see Section 9.4.2) have well-developed and formalized processes for responding to inquiries and issuing interpretations, others either have none or they are not well publicized. Because in some cases the SDO and the user community are well known to each other, with some standards it is common for a person with a question simply to call a member of the SDO and solicit an opinion or an explanation of code issues. An interpretation obtained in this manner is actually only an informed opinion from someone who is on the code committee, and cannot be expected to bind the SDO. This same approach is often used for organizational internal standards, where often there is a designated "owner" of a standard with some degree of administrative authority in addition to being, if not the writer, at least one of the writers of the standard. In this case an informal interpretation may, at least sometimes, be expected to carry more weight.

9.4.2   Formal Interpretations

9.4.2.1   Interpretations of Voluntary Consensus Standards

Committees in charge of standards often receive inquiries from users requesting interpretation of various aspects of the standard due to ambiguous language or lack of familiarity with the standard on the part of the user. The ASME BPVC committee publishes, at regular intervals, interpretations clarifying the intent of the code. The intent of the interpretations is

- to assist users to apply the code rules properly by clarifying ambiguous language;
- not to provide consulting advice;
- not to change or provide relief from requirements of the code.

One drawback of interpretations is that they may be applicable only to a specific edition of a standard. Some interpretations become obsolete when a new edition of a standard is published with corrections to the ambiguity addressed in a given interpretation.

An SDO with a well-developed procedure for processing inquiries and interpretations is ASME. The Foreword to Section VIII, Division 1 of the ASME BPVC includes the following text:

> *The committee meets regularly to consider revisions of the rules, new rules as dictated by technological development, Code Cases, and requests for interpretations. Only the Committee has the authority to provide official interpretations of this Code. Requests for revisions, new rules, Code Cases, or interpretations shall be addressed to the Secretary in writing and shall give full particulars in order to receive consideration and action (see Submittal of Technical Inquiries to the Boiler and Pressure Vessel Standards Committees) ...*

The section on Submittal of Technical Inquiries to the ASME Boiler and Pressure Vessel Standards Committees clarifies as follows:

> *Code Interpretations provide clarification of the meaning of existing rules in the Code, and are also presented in question and reply format. Interpretations do not introduce new requirements. In cases where existing Code text does not fully convey the meaning that was intended, and revision of the rules is required to support an interpretation, an Intent Interpretation will be issued and the Code will be revised.*

This section also provides requirements for submittal of requests for interpretation, with details of what information is required in order for an inquiry to be considered.

Not all organizations are as well organized as this for processing inquiries. In part, this is a reflection of the type and content of the ASME BPVC, the fact that this code has been around since nearly the turn of the last century, and the values of the organization. ASME recognizes the respect accorded the ASME BPVC and other ASME standards. It recognizes the importance of users being able to understand their standards and to receive authoritative interpretations if needed. Thus, it has a policy giving priority at its meetings to responding to inquiries. First on the agenda at every meeting of a code committee, after administrative burdens are taken care of, is responding to inquiries. Not every inquiry can be answered at the first meeting after it is received, but every reasonable attempt is made to do so.

Conversely, while an inquiry is sometimes necessary and SDOs typically try to give them priority, inquirers should also be aware of several issues related to inquiries: First, the nature of SDOs is such that they generally meet two to four times a year. Thus, even with the best of intentions regarding schedule, responses frequently take a year or more. Second, the response may be negative with respect to what the inquirer wants to do. Waiting 2 years for an answer, then finding out that that the answer is "No" will never be a positive development for a project. And, third, while some committees seem to provide particularly useful interpretations (e.g., ASME B16.34 Valves – Flanged, Threaded, and Welding End interpretations are particularly informative), others, being particularly careful to avoid acting as consultants, provide interpretations that are sometimes difficult to decipher.

9.4.2.2 Interpretation of Governmental Regulations

As with VCS, jurisdictional regulations sometimes raise questions as to their exact meaning or application, and similar processes may be used to provide interpretations. A letter can typically be sent to a jurisdiction, and a formal response be expected. In the case of OSHA, the letter is addressed to the Director, Directorate of Enforcement Programs. A response is made directly to the enquiring organization in a formal "Letter of Interpretation," which is also published on the OSHA website (https://www.osha.gov). Hundreds of such interpretations have been provided, addressing nearly every major section of the regulations. A similar approach is used by the Department of Transportation, with responses also published for general usage.

"On the spot" interpretations are also sometimes issued. It is usual for building departments, for example, to provide interpretations of requirements directly to inquirers standing across the counter with a set of plans for construction or modification of a house or other structure. In this case, the documentation of the interpretation is usually just the approval of the plans, and there is no formal publication.

9.5 Tailoring

Tailoring is the fitting of a standard to the particular needs, values, or priorities of a jurisdiction, organization, or program. This can involve addition, removal, or modification of requirements. It can occur when requirements are costly to implement, when a new technology is not yet readily available, or when the organization decides that certain requirements are not applicable to its particular needs, or do not make sense for it. As discussed below, tailoring is not without limits, as some standards are made mandatory by regulations, or are regulations themselves, and thus have the force of law.

9.5.1   Nonmandatory Standards Tailoring

An organization has the greatest flexibility for tailoring when a standard is not mandatory. If it simply chooses to follow a standard, then to some extent it can simply choose not to. Similarly, its own internal standards are written as they are because it best serves the organization, and when the needs of the organization change, the internal standards can be updated to meet those needs. Many governmental agencies have internal standards, and have a policy of revising, updating, or withdrawing them on a regular schedule. It must be kept in mind that tailoring is typically used for a somewhat different purpose than a waiver, in that it fits the standard to the specific needs of the organization, rather than just making an exception for a particular case.

9.5.2   Legislative Tailoring

Legislative tailoring is typically done during the adoption process. If a jurisdiction determines that certain requirements of a standard have not demonstrated their value, it may choose to take exception to those particular requirements, while adopting the balance of the document. Similarly, the jurisdiction may choose to adopt the standard either with changes or additions to the published standard. Building departments sometimes produce standard designs to supplement the Uniform Building Code in order to facilitate certain aspects of construction, then allow a simplified permitting process if those designs are followed (decks, retaining walls below a certain height, and finished basements are some areas where this is practiced). When legislative tailoring is performed, the effect is to make the standard mandatory, but with the specified changes, giving the tailored standard the effect of law.

9.5.3   Governmental Agency Supplemental Standards

The regulations of the OSHA provide prescriptive requirements for worker safety in many areas, and top level requirements in others. Various portions of the OSHA regulations address different areas of employment, including general industry, construction, shipyard employment, and safety of federal workers. If a governmental agency employs workers in an area for which there are no specific applicable OSHA safety regulations, then the "general duty clause" applies and the agency must develop safety standards for their work in that area and submit them for approval in accordance with 29 CFR 1910.18 (There is no direct parallel for this for non-federal agency employers, as such employers are simply required to address the issue in their safety plans, but are not obliged to submit their resolution to OSHA.).

9.6   Waivers and Variances

9.6.1   Waivers of Corporate Standards

While a corporate standard is in effect a policy of the company, and the company can therefore decide whether or not to follow it, there are practical reasons for use of a formalized process if it is necessary to deviate from full compliance. Corporate standards are often put in place not only to ensure rigor in procedures, but also to give customers confidence in the quality of company products and processes, or as implementations of government regulations. An organization will want to make certain that waiving a policy will not result in either a substandard product or failure to comply with a government regulation, and the formalized waiver process is used to ensure that the consequences of a waiver have been recognized and adequately considered prior to approving the waiver. One undocumented deviation can lead to another, and if care is not taken, there may soon be effectively no policy at all.

Waivers from a corporate standard are typically utilized when a material, product, or process does not meet the standard requirements but can still be used safely and effectively. These waivers are sometimes approved by the person in charge of the particular corporate standard, but often companies use a Material Review Board (MRB) process by which products that would otherwise be considered substandard and be rejected are evaluated by one or more specialists in order to determine adequacy for use or sale. The MRB will consider the likelihood and extent of potential consequences of the deviation from the standard. They are usually most concerned with form, fit, and function. A positive disposition will normally require a positive evaluation in each of these areas. If the product basically looks the same (form), the interfaces are the same (fit), and it works just as if it did not contain the variation from the standard, it will usually be accepted.

Although money is always a consideration, no company should make a short-sighted decision to accept a substandard component that will soon require replacement at company expense. Even worse would be a failure resulting in injuries, deaths, or even just property damage or productivity loss. A company is continuously aware of its reputation and will be reluctant to risk it in order to save what is in many cases a relatively small amount of money or time. Even a noncompliance with an internal or a nonmandatory policy that is accepted by a company can result in finger pointing and blame in the event of a mishap.

If a corporate standard is an implementation of a jurisdictional standard such as an OSHA regulation, then the substance of the waiver must be evaluated for compliance with that jurisdictional standard. If the alternate to compliance with the corporate standard is not in compliance with the regulation that it is intended to implement, then a waiver of the jurisdictional standard must be sought as well.

9.6.2 Code Cases

Standards are normally updated at specified intervals. However, instances arise where a component, a procedure, or a product may not fit exactly the requirements of the current controlling standard, and yet in the considered opinion of the governing code committee the proposed approach appears reasonable. Sometimes, there are no existing rules to cover the situation, or there may be rules, but because the particular issue simply had not been considered or allowed for, the rules may not allow it. There is a need for reasonable flexibility but a committee may not be ready to change the code, or the need is urgent and the normal course of changing the code may take too long. A code case (the expression used by ASME for this type of relief) is a way of providing some flexibility. It allows new approaches to be tried out without a permanent commitment on the part of the SDO, either while the matter is studied further or just to allow time to assess the need for and the effectiveness of the approach. Development of a code case generally allows the submitting organization and any other to use that approach, although some jurisdictions do not automatically accept them. The use of a code case also provides a means of documenting and accepting interim changes prior to the formal scheduled revision of a standard.

Accordingly, many standards have provisions that accommodate such situations. In the ASME BPVC, this provision is accomplished by issuing code cases. These code cases are written at the request of an inquirer. They cover a specific material, product, analytical method, or fabrication process. Generally, the code cases are

- limited in scope;
- intended to be temporary until the standard is updated or the need for the code case ceases to exist;
- less restrictive (though generally no less safe) than the rules in the body of the standard;
- published as a trial for a new analytical procedure or methodology. Comments received from the public will determine whether or not to revise the code case, incorporate it into the body of the code, or eventually annul it;
- limited in duration. Rules of the Boiler and Pressure Vessel Committee require it either to annul the code case after a period of time or to incorporate it into the body of the code.

Code cases in the BPVC are approved by the same consensus process used to approve other code items. They are published by ASME in two separate documents, one for nuclear components and the other for nonnuclear components.

Because code cases do not typically follow the same schedule as code revisions, they fall outside the normal approval process of jurisdictions that may require use of the code. As a result, many jurisdictions do not accept them in their laws as part of the ASME BPVC, or they require individual approval by the authorized inspector or the jurisdiction itself. Accordingly, the designer has to be aware of the laws in the jurisdiction where the vessel is installed.

One source of information regarding pressure vessel laws in various jurisdictions is the National Board of Boiler and Pressure Vessel Inspectors in Columbus, Ohio. Its website (https://www.nationalboard.org/) lists National Board members and Authorized Inspection Agencies, either of which should be able to provide assistance in various locations in the United States and Canada. If a code case is not accepted by a jurisdiction, then in order to apply it legally a company must seek and obtain the approval of the jurisdiction. This is separate from gaining approval of the applicable code committee.

9.6.3   Variances from Jurisdictional Standards

Jurisdictional standards are regulations, so they carry with them the force of law and associated sanctions for their violation. The governmental agencies promulgating these regulations (e.g., OSHA, the DOT, and local building departments) recognize this fact and their regulations include allowance for variances. The OSHA regulations will be used here as representative of the types of provisions that are made. While the requirements remain mostly the same, OSHA distinguishes between the federal government and other employers. For employers other than the federal government, relief from OSHA regulations is referred to as a variance, while federal government agencies are required to develop Alternate Standards if they wish to deviate from the regulation.

Jurisdictions occasionally permit variances to existing standards in cases where these standards do not directly apply to a given product, in cases where part of the standard is deemed inappropriate by the jurisdiction, or when the variance requester demonstrates an alternate product or methodology that effectively accomplishes the goals of the standard. The following example illustrates this situation:

> *State laws generally require pressure vessels to comply with the ASME PBVC in order to be operated in the state. In those cases in which it is for some reason impossible to comply with the BPVC but in which the owner can demonstrate an equivalent level of safety, it is usual for the State to accept the vessel as a "State Special." This makes it legal for operation within the jurisdiction of that state, but if such a vessel is moved to another state, it will be necessary to seek another "State Special" authorization.*

> *Similarly, pressure vessels constructed in accordance with the ASME BPVC but with a code case may need explicit approval of*

the jurisdiction if the jurisdiction has not already accepted that code case.

On occasion, pressure vessels for a proprietary process are built in a foreign country in accordance with that country's pressure vessel standards. The vessels may then be incorporated in a product and shipped to the location of use. In such instances, the state frequently grants variances from the requirement to comply with the ASME BPVC and accepts the foreign code of construction for this particular vessel. It should be noted that in these cases the foreign code is only accepted because it has a level of credibility similar to that of the ASME BPVC, although it differs in details and application.

9.6.3.1   General Industry Variances from OSHA Standards

Temporary Variances allow time for an employer to meet a newly published requirement. This is provided for in the Code of Federal Regulations, 29 CFR 1905.10, and allows additional time when changes to a facility cannot be made quickly enough to comply with the specified effective date of the regulation.

Permanent Variances allow an employer to use alternate means, on a permanent basis, if the employer can demonstrate that the workplace is at least as safe and healthful as would be provided by the standard from which the variance is sought (29 CFR 1905.11).

National Defense variances allow reasonable variances, tolerances, or exemptions to avoid serious impairment of the national defense (29 CFR 1905.12).

Experimental Variances are provided in Section 6(b)6(C) of the OSH Act of 1970, and authorize an employer to demonstrate new practices. *Note the similarity between an Experimental Variance to the OSHA Regulations and a Code Case to one of the ASME Codes.*

An interim order may be granted during review of an application for either a temporary or a permanent variance.

While this section describes five different types of variances allowed by the OSHA regulations, it should be noted that each type still requires demonstration that the employer will maintain a workplace at least as safe as that provided for by the regulations.

9.6.3.2   Governmental Agency Alternate Standards

Essentially the same as a permanent variance for industry, federal government agencies may apply for an alternate standard to the OSHA regulations, provided they offer a workplace at least as safe as would be provided by compliance with the OSH Act. Specific details of what must be submitted are provided in 29 CFR 1960.17.

9.7 Summary

With standards in existence addressing almost all aspects of engineering, it is inevitable that there will at times be a need for interpretation of, or relief from, standards. Several things should be considered before relief is requested:

1. Is there really a need for relief? This involves a careful reading of the standard to determine precisely what it means, and whether the supposed activity or product is really out of compliance with the standard. Waivers have sometimes been requested for compliant activities.
2. Is a request for interpretation in order rather than, or before, a waiver request?
3. Is there reason to expect that the request for relief will be approved? Is the product as safe and effective as a compliant one?
4. What are the broader implications of the noncompliance? Will it result in a lesser product? Potential liability or damage to a company's reputation? Possible failure to fit with compliant mating products at a later time?

If there is a valid reason to request relief, then such a request can make sense and there is a good chance that it will be granted.

9.8 Case Study

A nationally known company produces door hardware for the residential market, manufacturing privacy knobs, entry knobs, deadbolts, and other products. It has an extensive set of internal standards and specifications built up over years of experience in the field. A common lock barrel consists of a cylinder with a series of precisely sized and located holes in which precision hardened pins slide into alignment when the key is inserted. This company has a reputation for quality built up over generations through careful attention to product quality as well as thoroughly addressing every comment or complaint that they receive from a customer.

A question about one product line sometimes requiring jiggling the key in order to get the lock to function was brought to the attention of the quality department, resulting in a careful evaluation of a number of locks. It was determined that the length of one size of pins was slightly out of tolerance and that the jiggling of the key allowed their ends to "adjust" into alignment, permitting the lock to function. It did not appear that any of the locks was likely to fail as a result of this nonconformity. Any continued use of the parts would require processing of an internal (company) waiver.

Various options are available to the company, including to

1. accept the nonconforming parts and continue to sell the product as is;
2. correct the nonconformance, but leave those locks already sold on the market;
3. reject all the nonconforming components and conduct a complete recall of the product.

Discuss the advantages and disadvantages of each approach.

10

Characteristics of a Good Standard

10.1 Introduction

In Chapter 2, it is pointed out that standards are generally written to achieve safety and reliability, to reduce cost and increase flexibility, to promote business, and to help society function. A well-written standard will accomplish these goals effectively. A poorly written one will accomplish them less well, and may even impede them. This chapter identifies those characteristics that make for an effective standard. These characteristics help in selection when there is a choice of standards, development when a new standard is needed, and effectively supplementing an existing standard with proper text in a specification, statement of work, or other contract documents.

10.2 Clarity and Understanding

In order to achieve its intent, a standard must first be understood. This seems obvious, but since this discussion is related particularly to engineering standards, it may be more difficult to accomplish than it first appears. The nature of many engineering standards is technical, sometimes highly so. While not everyone will be able to understand fully the content of a technical document, a standard should be written, and provided with sufficient definitions so that a technically competent user can understand it. Further, it should be written such that a person not well versed in the field, but taking time to read it carefully, can comprehend the general concepts and requirements, if not the technical details. In some cases this can be a challenge, as when a standard of limited scope in a specialized

Primer on Engineering Standards: Expanded Textbook Edition, First Edition.
Owen R. Greulich and Maan H. Jawad.
© 2018, The American Society of Mechanical Engineers (ASME), 2 Park Avenue, New York, NY, 10016, USA (www.asme.org). Published 2018 by John Wiley & Sons Ltd.

field makes reference to and has requirements for compliance with one or more similarly technical and specialized documents.

If a standard assigns responsibilities to a person or organization, the responsible party should be clearly designated. If there are special qualifications that go with assigned duties, these should be clearly indicated. While a number of standards developing organizations make a practice of using "shall" for requirements, "should" for recommendations, "will" for expected consequences, and "may" for permissions, some organizations have other conventions. Whatever terms are selected must be clearly defined and assiduously used.

Many organizations lean toward performance standards over prescriptive ones, the idea being that this allows the performing organization the greatest flexibility to implement better or cheaper ideas. Also, standards are written to aid in accomplishing something, and a well-written performance standard is closer to defining what is needed, rather than a solution that may not be the best one. On the other hand, it is often difficult to define every requirement. If every requirement has not been well thought out and documented, then prescribing a known successful solution is often safer than trying to identify every detail needed for an effective performance standard.

10.3 Scope

The scope of a standard ranges from very narrow and specialized to very broad. Whatever be the case, it should be clear and well defined. The following two examples illustrate this:

Example 1: ASME SA-20, Specification for General Requirements for Steel Plates for Pressure Vessels, consisting of only 41 pages of the ASME BPVC Section IIA, includes common requirements that apply to 33 steel plate specifications for fabrication of pressure vessels. It references 50 standards, including the 33 plate specifications as well as 17 standards related to terminology, testing and inspection, welding, marking, and handling. This standard defines within itself the requirements for how plate should be ordered and manufactured, test methods and reporting and retests, quality and rejection criteria, tolerances, marking, packaging requirements, etc. It leaves to the individual plate specification the actual chemical and physical properties, whether certain tests are required, and it leaves to the purchaser identification of any supplementary requirements, although it specifies many details for the case where supplementary requirements are specified. This standard has a fairly narrow scope.

Example 2: ASME B31.3, Process Piping (originally the Chemical and Petroleum Refinery Piping code), at over 500 pages, was developed "considering piping typically found in petroleum refineries; chemical, pharmaceutical, textile, paper, semiconductor, and cryogenic plants and related processing plants and terminals." It applies to all fluids, but has clearly defined exclusions. It references hundreds of other standards and hundreds more by references within those. It includes by reference the ASME SA-20 specification referred to above. This standard has a wide scope.

The difference in scope between these two standards is immense, but a study of each document reveals that the scope is clearly defined, and that each document addresses its scope without straying beyond. Further, across the scope of each document, the depth of treatment is relatively consistent. ASME SA-20 covers only the general requirements for certain steel plate material, while ASME B31.3 addresses the production of piping systems for a wide variety of applications and using a wide range of product forms. In order to maintain a consistent level of detail and to avoid a document running to thousands of pages, B31.3 makes reference to plate, bar, forging, pipe and tube, and other standards. These standards, in turn, reference others, including ASME SA-20, and SA-20, in its turn, requires compliance with other standards dealing with the rolling process, nondestructive evaluation, etc.

This approach allows a fairly concise document addressing the specific concerns of those involved with production of steel plates (in the case of ASME SA-20) or production of process piping (ASME B31.3). The details of how to perform NDE of the steel plates in the case of SA-20, or, tolerances, NDE, etc. of those plates, in the case of B31.3, are included in reference standards that address the specific needs of those involved in those particular specialties.

By constructing what might be referred to as a system of standards, the creators of ASME B31.3 have produced an efficient document that can be used in conjunction with whatever reference standards are appropriate for a particular situation (and the particular business that a company might choose to pursue). This provides an efficient way of managing a vast number of requirements.

10.4 Terminology

Use of specialized terminology, and of terms with a meaning different in the context of the document than in common speech, should be clarified. In this respect the document should be able to stand on its own. Specialized terms and expressions may be unfamiliar to the general reader, yet may need to be used in order to convey accurately the intent of the document. The process of standard development typically includes discussions of which terms require definitions, and which ones are sufficiently understood from the common use of language or from definitions understood in the engineering field in general.

10.5 Structure and Organization

The structure of a standard is dependent in part on its content, but just as there is more than one way of structuring a corporation, there is more than one way of arranging information.

To provide for their efficient use, engineering standards are most often constructed in a modular manner, whether internal to a single standard or as what might be considered a system of standards. A top level standard may refer to a number of other standards that help in the definition of an overall product. Although some repetition may occur, this approach minimizes the need to repeat common requirements.

For a standard addressing the design and fabrication of a product, it is usual to provide a design section distinct from those sections addressing fabrication, inspection, testing, qualification, etc. It would, however, be possible to place the inspection requirements in a standard next to the aspect of fabrication to which each applies. Most standards follow the former organization. This approach seems to work well, although its use is partly for historical reasons (this is the way many standards have been written for the last century) and partly because so many of the supporting standards fit better with a document that is arranged this way.

For ease of use, it is usual to keep related information together in a standard. Consistent with the strategy, the standard developer should strive to place all text, figures, tables, and equations for a given topic, say, earthquake analysis, in one section rather than spread it out over many chapters. An organization that requires the user to flip back and forth between chapters to extract information makes the use of that standard inefficient for the engineer and increases the likelihood of errors.

An example of effective placement of information is found in Mandatory Appendix 2, Rules for Bolted Flange Connections with Ring Type Gaskets, of the ASME BPVC Section VIII, Division 1. Everything from bolt loading to gasket type to figures for use in calculating stress, is found in this single appendix, generally in the order in which it will be used by the engineer.

Once a structure and organization is selected, it is important that it be followed, and whatever structure is followed, it should be logical, such that once a person becomes familiar with the standard he/she will find information where it is expected. There should be no need to go on an extended search for requirements that clearly "must be there somewhere." At least one standard exists in which not all the design requirements are found in the design section (some are in the testing section), and not all the testing requirements are found in the testing section (some are in the design section). This leads to confusion, much searching around, and the possibility that requirements will be missed by a user who is not thoroughly acquainted with the document. There are plans to correct this situation in the next revision of this particular standard, but a little more thought given to the structure of the standard and placement of information within it could have avoided the problem.

10.6   Consistency

Just as it is important to follow the selected structure of a standard, it is important that the standard be consistent. In developing a standard, particularly a complex one of any length, it is not unusual to address different aspects of an item in different places or sections of the document. Because these sections may be separated by a large amount of text, and may in fact be written by different authors because of the way work is divided, care must be taken to examine the document as a whole to ensure consistency. Standards developing organizations often find themselves in the position of addressing inquiries inspired by apparent inconsistencies, and at times even making changes to correct such inconsistencies. A particularly helpful aspect of organization of the ASME B31 piping codes is that each follows the same structure, so that the requirements for a particular aspect of piping design or fabrication are found in the same paragraph in each of the codes, differing only in the first digit of the paragraph number, which reflects the ".X" designation of the respective standards (e.g., for ASME B31.3, paragraph numbers begin with 3).

10.7   References to Other Standards

Most engineering standards will require references to other documents and standards. This need should not prevent a standard being consistent, unambiguous, and fully covering its stated scope. As can be seen from the examples above, a standard need not explain every detail. Further, standards are designed to be used together to achieve their goals. When other standards addressing a particular aspect of the scope of a standard already exist, those other standards can help fill out the scope if properly referenced. This approach avoids both redundancy and contradictions. It saves development time and minimizes the need for deciding among multiple standards covering the same topic; and finally, it allows those already familiar with the referenced standard to proceed without the need to study another document.

10.8   Attention to Details

It should perhaps not need mentioning, but the spelling, grammar and syntax, references and cross references, and every other detail of a standard should be thoroughly checked and rechecked. If "shall" has been selected as the term for an imperative, for example, then a word search should be made for every other likely term that may have slipped in, so that each can be corrected if needed. Many people will be depending on the quality of the product.

10.9   Supplementing a Standard

When a standard is cited or otherwise put into effect by an organization, there may be a need either to provide additional guidance or to take exception to certain aspects. Whatever contractual terms are used should hew to the same level of quality that is expected of the standard itself. There is little value in having a clear, well-thought-out, and documented standard which is not clearly invoked, leaving it unclear as to what the actual requirements for a project or task are.

10.10   Timeliness

Often a standard is released for distribution prior to it being completely finished owing to time constraints. This results in confusion on the part of the user and the end result is a reluctance of the targeted audience to use the standard. When this situation occurs, the organization in charge of the standard scrambles to rewrite the standard, filling in the voids and restructuring it as needed. The end result is that it takes longer to release an incomplete standard and fix it than to wait and release a complete standard. Sufficient schedule time must be allowed in order to avoid this situation.

10.11   Sample Standard Structure

For a standard to be widely accepted, it must be easy to use. Accordingly, good organization of a standard is of an utmost importance and the topics must flow smoothly in order to avoid having the user go back and forth in the document. The following is a summary of the essential elements of a standard:

1. Table of Contents. A detailed Table of Contents is very helpful in pinpointing the topic of interest. It, or the index, is normally the first place the user visits prior to using the Standard.
2. Foreword. The foreword normally explains the reason for developing the Standard and the process by which it is developed.
3. Scope/Preamble. The scope or preamble defines the applicability of the Standard as well as its limitations, exceptions, and assumptions. It helps the user understand the applicability of the Standard. Sometimes there are two or more standards that cover the same subject but the applicability is quite different. An example of this situation includes the repair standard for pressure vessels published by the National Board (NBIC) and that published by ASME (ASME FFS-1). They both cover the same subject but are vastly different in their applicability. The NBIC has general repair rules but references other standards for design and analytical parameters. ASME FFS-1 on the other hand has detailed design and technical parameters and methodology for handling repairs. Accordingly, the technical ability of the NBIC code user might

be different from the technical ability of the ASME FFS-1 user even though the end result tends to be the same.

4. Terms, Definitions, Acronyms. There is no need for definition of commonly used and understood terminology, but definitions unique to the standard should be provided. This section is sometimes found early in the document, or sometimes as an appendix or a glossary at the end.

5. Reference documents. The need for this section varies, as does its placement, but most often it is not only helpful but also necessary.

6. Topics. Sequence of the topics in a standard varies greatly depending on the type of standard. The following list of topics might be typical for a design standard for fabricated products, and is not intended to represent an exhaustive list:

   a. Design conditions
      i. Dimensional requirements (volumetric minimum and/or maximum, envelope dimensions, etc.)
      ii. Materials (if specified)
      iii. Loads (force, pressure, moment, etc.)
      iv. Life (cycles, years, etc.)
      v. Environments (internal and external, as applicable)
      vi. Cleanliness, etc.

   b. Design criteria (may range from fairly prescriptive to fairly broad)
      i. Safety factors or allowable stresses
      ii. Stress analysis
      iii. Fatigue analysis

   c. Materials (might specify acceptable materials, define characteristics of materials to be accepted, or define processes for determining material properties to be used; may not be applicable in some process standards)

   d. Fabrication
      i. Forming
      ii. Welding
      iii. Post weld heat treatment
      iv. Surface finish

   e. Inspection
      i. Dimensional
      ii. Visual
      iii. NDT

   f. Testing
      i. Qualification testing (used to qualify a design)
      ii. Acceptance testing (less stringent than qualification testing, used to determine acceptability of an individual article)

   g. In-service inspection

h. Refurbishment and/or repairs

i. Documentation

7. Appendices. These typically provide information that can stand separate from the rest of the document. They may include information that applies only in special cases or applications, or may provide guidance on topics that are not the main thrust of the document (considerations in performing a pneumatic test on a piping system, for example, in order to ensure a safe and effective test).

8. Index. A detailed index is an important tool in a Standard for helping the users find specific topics and details.

10.12   Summary

The quality of a standard is critical to its usefulness. The characteristics listed above are the basics needed to ensure a useful and effective standard. Each person or group developing a standard should give careful thought to every aspect of the product being produced, and then follow through to ensure that their standard is of a quality such that they, and any other users, will find it effective and easy to use.

10.13   Case Studies

Case Study 1

Using a bicycle to go back and forth to work is becoming more prevalent throughout various metropolitan areas in the industrial world. It is a good exercise and saves natural energy resources. Write the outline of a standard for minimum bicycle requirements when used on major streets and highways with automobile traffic. Outline a standard defining the mechanical, safety, and electrical requirements.

Case Study 2

Select a product or process from your field of interest, and (1) determine what type of standard is most appropriate for it and (2) provide an outline of that standard.

11

Getting Involved in Standards Development

11.1 Introduction

The many thousands of standards currently in existence require the work of many more thousands of people dedicated to their development and maintenance. While administrative aspects are often taken care of by the SDO, the technical work is typically performed by highly qualified experts in their respective fields, most often funded by their employers who are affected by the standard. There are many reasons to be involved in standards development, and many opportunities for interested parties to contribute significantly to society by doing so.

11.2 Reasons to Get Involved

Interested parties see an opportunity to get involved, although the reasons vary from person to person and organization to organization. Some reasons are discussed here.

11.2.1 Influence the Process and the Product

Many organizations that are influenced by standards want to have a say in what those standards look like, what they cover, and what they require. Companies recognize that standards can be beneficial to them, but also that a poorly written standard, one with incorrect, insufficient, or excessive requirements can be

Primer on Engineering Standards: Expanded Textbook Edition, First Edition.
Owen R. Greulich and Maan H. Jawad.
© 2018, The American Society of Mechanical Engineers (ASME), 2 Park Avenue, New York, NY, 10016, USA (www.asme.org). Published 2018 by John Wiley & Sons Ltd.

detrimental to them and to their industry. By contributing skills to the production of a standard, a company can ensure that its needs and views are considered in the development process, that the standard is reflective of current state of the art in that field, and that the level of detail and the requirements in the standard are such that the standard is a help and not a hindrance. One of the easiest ways to get what you want is to volunteer to help produce it.

Sometimes, a person is assigned by his or her organization to participate in one or more SDOs as a way of protecting the interests of the organization. Some companies fund engineers who work almost entirely on the development and updating of standards that affect aspects of their industry.

11.2.2 Opportunity to Learn

Working on the development or updating of a standard requires a thorough understanding of the product or process in question, or a willingness to develop that understanding. It provides one of the best ways to further the understanding of the details, and of the standard as a whole. It demands an appreciation of the particular requirements being written, as well as their relation to other requirements. Details need to be worked out. The text must be clear and unambiguous, sometimes requiring a number of iterations to achieve consensus. What is clear to one person may not be clear to another, or two people may agree that something is clear, but have different understandings of what it (clearly) means. The standard needs to be reviewed to ensure the accuracy of cross references, and consistency between sections and with other standards that are referenced.

Working through all of these details results in a familiarity with the standard that is difficult to achieve in any other way. Thus, those working on a standard typically come to understand it better than nearly anyone else, including an appreciation of its history, the background and technical basis for requirements, and relationships of that standard with other requirements and standards.

11.2.3 Credibility

Usually more people and organizations use and depend upon a standard than are involved in its development. Being on a standards committee is generally recognized as reflecting a higher level of understanding and qualification in the field than that of the general user. While some standards committees are easy to join, and almost all are open to public attendance, taking time to meet with other specialists in the field and to work through technical issues and the details of wording results in a familiarity with the standard and an understanding of it beyond what the casual user will achieve. Volunteers are generally respected for their involvement in standards developing organizations.

11.2.4   Personal Satisfaction

Helping to develop a product that is important, and knowing that one's efforts have contributed to the quality of that product and have wide ranging effects, can be very rewarding. This satisfaction is manifested every time the person who has worked on the standard uses it or works with it, including helping others to understand and use it.

11.2.5   Networking and Career Benefits

The process of developing a standard typically brings together a number of people, including many of those with the greatest amount of theoretical and practical knowledge in the field. There are often subspecialties within the field, and the various parties involved have the chance to share ideas and learn from each other.

For example, the AIAA Standard for Space Systems – Composite Overwrapped Pressure Vessels (COPVs), includes requirements for design, fabrication, testing, inspection, operation, and maintenance of COPVs. It is applicable to COPVs used for a wide range of pressurized applications, including hazardous and nonhazardous, liquid, and gas, used on both spacecraft and launch vehicles. While a number of people have spent much of their careers working in this area, even they have specialized and do not have the full range of knowledge of these products. Some are materials experts specializing in material physical properties such as strength and fracture characteristics, while others are particularly concerned with materials compatibility. Some are heavily involved in the fabrication process, including how liners are produced or how resin is applied to the fibers and how the fibers are applied to the liners. Others specialize in testing and inspecting the COPVs. Getting involved in developing a standard gives the participant the opportunity to interact with others who have different specialties within the same sphere of work.

By meeting periodically to develop and to update the standard for COPVs, those most interested in these products across the industry have a chance to become acquainted and work with each other.

In addition, whether employed by someone else or working for oneself, the knowledge that has been acquired working on codes and standards will be useful. Being a member of a standards committee can be a definite selling point for consultants, as can having such a member in one's employ. An engineer seeking the next opportunity for advancement, whether inside a company or in moving to another, is likely to gain credibility through having helped develop a standard that the company regularly employs. And finally, the networking opportunities of working on a standards committee can be tremendous. It has been recognized for a number of years that a significant percentage of jobs are filled without formal search or advertisement. Word of mouth is a highly effective means of making contacts.

11.3   Opportunities for Involvement in Standards

11.3.1   Company Standards

Most large organizations utilize both internally developed and external standards. A person who is recognized within an organization as holding a reasonable level of expertise in a given area can often get involved in the improvement of standards simply by providing ideas or volunteering to work on a particular aspect of a company policy or other document. Companies depend on their standards as a means of ensuring consistency of products, to promote efficiency, limit liability, and for a variety of other benefits. Work on these documents, whether their development or maintenance, is often considered to be a less than ideal assignment, yet it often provides a way for an employee to better understand the company and how it does things, to interact with management, and to make real improvements in a company's products or processes. It has a further potential to lead to assignments on voluntary consensus standards as well.

11.3.2   Interest Group Standards

Work on interest group standards such as those of the Expansion Joint Manufacturers Association or the Tubular Exchanger Manufacturers Association may be considered prestigious assignments, yet if an employee has proved willing and able to work on company standards, there may be opportunities here as well. The more qualified the employee, the better the chance of involvement. That said, some standards in this realm have been developed as much for marketing purposes as for technical ones, and once in existence they may remain fairly static for long periods of time.

11.3.3   Voluntary Consensus Standards (VCS)

The vast numbers of VCS with requirements for openness and for a range of representation on their developing organizations offer the greatest number of opportunities for standards participation. Thousands of standards require tens of thousands of people to develop and maintain them. In addition, because most of them follow the requirements of the American National Standards Institute for VCS, opportunities are widely available for interested parties. Thus a person or organization wishing to participate generally can do so. Even a person who is not a full committee member can have a significant influence on the development of a standard. In addition, many of these standards are the very same ones that provide the greatest prestige, opportunities for learning, and networking opportunities. Yet, they often lack willing workers who can dedicate sufficient time and thought to keep the standards current.

11.3.4   Jurisdictional Standards

Opportunities for direct participation in development of jurisdictional standards (also referred to as *regulations*) are often somewhat more limited. When technical standards are embodied in laws, they are often created by employees of the agency assigned to develop implementing regulations for particular laws. That said, there is still often a need for outside expertise, as it is difficult for a governmental agency to maintain all the talent needed to create effective technical regulations. Further, once the standards have been written, they must almost all go through a public review period. This is a period during which any interested party can provide comments and suggestions on the published document, and have them addressed, prior to it going into effect.

11.4   Selecting a Committee

Standards development is well worth doing and can be very rewarding. It is possible to be on a standards committee and do very little work, but the greatest benefits occur for the individuals most committed. For those committed to ensuring that a standard gets completed in a timely manner and that it is a quality document the task can be very consuming, both in time and energy.

If you plan to invest significant effort in something, it is worth taking a little time to make a wise selection regarding what you will work on and the work that you will do. This means researching the available opportunities and selecting a committee that will be rewarding to work on. The following sections provide some ideas about how to accomplish this.

11.4.1   Finding a Committee

Often finding an appropriate committee to join is as easy as noting what standard you most often use in your work. Or it may involve identifying a standard that you would like to understand better. It may revolve around identifying the committee members you would like to work with. The process of joining a subcommittee often involves visiting several to determine which offers the best overall fit.

If you have selected a particular standard you would like to get involved with, there is still often more to think about. Standards committees often have subcommittees that perform most of the work. For example, currently the American Society of Mechanical Engineers (ASME) publishes approximately 550 technical standards, developed and maintained by about 5000 technical experts. Hardly a week goes by without an ASME committee meeting, and it is common for these meetings to involve many subcommittees, addressing different aspects of the standard.

While being a member of a higher level committee allows one to vote, it is the subcommittees that generally develop the actual text that goes into the standard. The ASME B31.3 Process Piping Code Committee, for example, has subcommittees on materials, design, high pressure, and other areas, such that eight or ten subcommittees are actually meeting twice a year. Subcommittee meetings are normally concurrent, so a person is forced to decide which one to attend. If the member or organization has interests in multiple areas, it can be a juggling act to be at each meeting at the right time (the members understand this, however, and work to accommodate each other).

11.4.2 Making the Choice

Personal interest, employer interests, personal benefits, and opportunities to contribute all go into making the decision about which committee to join. Sometimes, the decision revolves around which committee needs help the most. Even if there is not a perfect match, however, many of the same benefits are achieved. And as implied above, a decision to join one committee or subcommittee does not prevent one from contributing to others on an intermittent basis or through the public comment period.

11.5 What Does It Require?

Participating in a standards developing organization requires a number of things. First is an interest and willingness to do so. If these are present, then the knowledge and abilities will follow, and most committees are willing to have a somewhat less experienced member if that person is willing to contribute to the less technical sides of the work while developing sufficient expertise for other aspects.

Often participating on a Standards Developing Organization (SDO) requires a certain amount of travel, with many such groups having face-to-face meetings one or more times per year. Travel involves time away from home and family, and it costs money. It is common for interested companies to cover the travel costs of their employees working with SDOs.

Being a successful member of an SDO requires judgment and a willingness to compromise. Different participants have disparate interests. A manufacturer may in some cases want as few constraints as possible, while hoping for the maximum in legal protection in case of a failed product, while a participant using the product produced to the standard may prefer to have a more stringent set of requirements. Each must recognize the needs and wants of the other participants. If this happens, then a reasonable balance can almost always be achieved.

Membership in some cases is achieved simply by asking and filling out a form. In other cases, it requires a vote of the membership. In this latter case, it is common for a person to attend several meetings and to work with the committee prior to attaining membership, as a means of demonstrating the ability and willingness to work. After several meetings the person is then voted in as an official member.

11.6 Summary

Participation as a member of an SDO provides a number of benefits both to society and to the participant. The opportunities to participate are many and various. Because involvement often requires significant commitment on the part of both the member and his/her organization, it is wise to consider carefully before deciding to join a particular committee. There should be a clear understanding between employees and their employers as to the responsibilities of the respective parties, including time off work to participate, time during work devoted to standards activity, cost of travel, what happens to workload in the worker's absence from his/her regular job, and whether the employee represents him/herself or the employer when working on the SDO. If these issues are worked out, and if the employee brings both capabilities and a willingness to work with others, then this participation can be rewarding for all concerned.

Acronyms

Acronym	Name
AASHTO	American Association of State Highway and Transportation Officials
ABMA	American Bearing Manufacturers Association
ABMA	American Boiler Manufacturers' Association
ACI	American Concrete Institute
ACMA	American Composites Manufacturers Association
AGA	American Gas Association
AGMA	American Gear Manufacturers Association
AHRI	Air-conditioning, Heating, and Refrigeration Institute
AIA	Aerospace Industries Association
AIAA	American Institute of Aeronautics and Astronautics
AISC	American Institute of Steel Construction
AISI	American Iron and Steel Institute
AITC	American Institute of Timber Construction
ANS	American National Standards
ANS	American Nuclear Society
ANSI	American National Standards Institute
API	American Petroleum Institute
ASA	Acoustical Society of America
ASABE	American Society of Agricultural and Biological Engineers
ASCE	American Society of Civil Engineers
ASHRAE	American Society of Heating, Refrigerating and Air-Conditioning Engineers

Primer on Engineering Standards: Expanded Textbook Edition, First Edition.
Owen R. Greulich and Maan H. Jawad.
© 2018, The American Society of Mechanical Engineers (ASME), 2 Park Avenue, New York, NY, 10016, USA (www.asme.org). Published 2018 by John Wiley & Sons Ltd.

Acronym	Name
ASME	American Society of Mechanical Engineers
ASNT	American Society for Nondestructive Testing
ASTM	American Society for Testing and Materials
ATIS	Alliance for Telecommunications Industry Solutions
AWC	American Wood Council
AWEA	American Wind Energy Association
AWS	American Welding Society
AWWA	American Water Works Association
BPVC	Boiler and Pressure Vessel Code
CGA	Compressed Gas Association
CISPI	Cast Iron Soil Pipe Institute
COPV	Composite Overwrapped Pressure Vessels
CPSC	Consumer Product Safety Commission
CSB	Chemical Safety Board
DOD	US Department of Defense
DOE	US Department of Energy
DOT	US Department of Transportation
EJMA	Expansion Joint Manufacturers Association
EPA	US Environmental Protection Agency
EPRI	Electrical Power Research Institute
FAA	US Federal Aviation Administration
FDA	US Food and Drug Administration
FFS	Fitness for Service
HI	Hydraulic Institute
HPVA	Hardwood Plywall and Veneer Association
HTRI	Heat Transfer Research Institute
IBC	International Building Code
ICC	International Code Council
IEEE	Institute of Electrical and Electronics Engineers
LCS	Limited Consensus Standards
MRB	Material Review Board
NACE	National Association of Corrosion Engineers
NASA	National Aeronautics and Space Administration
NBBI	National Board of Boiler and Pressure Vessel Inspectors
NBIC	National Board Inspection Code
NCS	National Consensus Standards
NEC	National Electric Code
NEI	Nuclear Energy Institute
NETA	InterNational Electrical Testing Association
NFPA	National Fire Protection Association
NIST	National Institute of Standards and Technology
NRC	US Nuclear Regulatory Commission

Acronym	Name
NRS	Nameplate, Records, and Stamping
NSF	National Science Foundation
NSTC	National Science and Technology Council
NTTAA	National Technology Transfer and Advancement Act
OMB	Office of Management and Budget
OSHA	US Occupational Safety and Health Administration
PCA	Portland Cement Association
PSM	Process Safety Management
SAE	Society of Automotive Engineers
SDO	Standards Developing Organization
TEMA	Tubular Exchanger Manufacturers Association
TMS	The Masonry Society
UL	Underwriters Laboratories
USB	Universal Service Buss
VCS	Voluntary Consensus Standards

Appendix A

Deciding Not to Use a Standard

A.1 Introduction

This book has dealt with the many aspects of standards, including their benefits. If a standard is available for an application, it is generally wise, and it is sometimes required, to use it. A number of benefits of doing so are outlined in Chapter 3. In spite of these benefits, there will be cases in which use of a standard may not be the most effective way to produce a particular product or perform a particular task. Before a decision is made not to use an available standard, however, the need, the consequences, and the mitigations for not complying with the standard should be carefully considered.

If a standard is required by law or regulation (e.g., OSHA, DOT, or FAA regulations) that fact should not be ignored. A decision not to comply with those regulations may well involve legal consequences and the jurisdictional organization should be consulted before moving forward. Even if the decision is technically justifiable, a waiver or other documentation may be needed in order to avoid regulatory consequences.

In Chapter 3, it was noted that use of a standard can provide benefits of improved reliability, interchangeability, reduced costs, confidence in the product, process, or design approach. In the event of a failure it may bolster the argument that the product was manufactured responsibly. These benefits should not be given up lightly.

Primer on Engineering Standards: Expanded Textbook Edition, First Edition.
Owen R. Greulich and Maan H. Jawad.
© 2018, The American Society of Mechanical Engineers (ASME), 2 Park Avenue, New York, NY, 10016, USA (www.asme.org). Published 2018 by John Wiley & Sons Ltd.

A.2   Reasons Not to Use a Standard

There are a number reasons why a standard may not be used, divisible into a few general categories.

- Lack of a standard.
- The most applicable standard is overly constraining.
- The product or process is unique and the standard is not a good match.
- The product or application is so basic as to be generally recognized as not requiring a standard.
- Cost.
- The product is patented. Hence, there is no need for a standard.

A.2.1   Lack of a Standard

There may not be an applicable standard, and while a company could choose then to develop its own standard, it might also decide that such an effort is not warranted because of limited need. Also, related to this is the situation in which there may be a standard that could be applied, but the match is not good. An example of a case in which there is no standard is the design of strakes in slender towers and chimneys for preventing the formation of wind vortex shedding, and hence vibration.

A.2.2   Overly Constraining

The most applicable standard may be too constraining. For example, the standard might specify acceptable materials, and those allowed may not meet the special needs of the customer, such as chemical compatibility, heat transfer characteristics, or strength, resulting in a product that might react with its contents, one that overheats because it cannot transfer energy out quickly enough, or weighs too much. This case might be represented by dimpled jackets on pressure or vacuum shells, for which the design is based on testing rather than using very conservative rules found in many international codes and standards. Another example is vessels constructed for use in high vacuum applications, for which the ASME Boiler and Pressure Vessel Code (BPVC) could be used, but for which certain important design details (e.g., intermittent welds on the outside of nozzles) are not accepted by that code.

A.2.3   Unique Product or Application

Sometimes, the uniqueness or the low volume of a particular product combined with a perceived lack of need makes the use of a standard unnecessary or undesirable. In this case, there may still be some standards that apply and are used for particular aspects (e.g., materials), but the overall product may be constructed without reference to a specific standard.

An example of such a case might be a pasta machine. The making of pasta may be considered by some to be more of an art than a science, with the thickness and width of individual pieces a matter of personal preference. There are many pasta machines on the market, with a wide range of capabilities, features, characteristics, and levels of quality. Electrically driven ones will almost certainly be listed with Underwriters Laboratory and meet certain standards for their drive and electrical systems. The steels used for the cutters may well be ASTM listed products. Safety requirements of the CPSC will need to be met. But the major features that actually produce the pasta are most likely not constructed to a standard. Indeed, there is probably no such standard, whether for home or industrial pasta machines.

Similarly, in process equipment there are many geometries and components that are not covered by standards. This is due to their limited use or limited applications. These include cylindrical shells with elliptical cross sections, seal-weld gaskets, and special configuration bolts.

A.2.4   Basic Services

Many services do not require compliance with a standard. These include tasks such as mowing a lawn or running a car wash.

A.2.5   Patented Products

Since a patented product is intended to be used by one company or its licensees, the need to develop a standard to have uniformity in design is not warranted.

A.3   Consequences of Not Using a Standard

The consequences of not using a standard can include the loss of any of the benefits of using it. Which ones apply will depend on the particular case in question. A product built for in-house use by a company's employees, for example, must meet any OSHA regulations, including referenced standards, but if there are no mandatory standards then the compliance aspect is not relevant. Similarly, in this case there is no need to develop credibility with the public or a customer base. Other aspects, however, such as confidence in the design and interchangeability of components may be lacking. If the same product were built for sale, then customer confidence and a potential credible defense in the case of a failed product would be factors.

A.4   Mitigations for Not Using a Standard

As previously noted, there are often distinct advantages to the user of a standard, as well as to the purchaser of the resulting products or services. Failure to work to a standard typically gives up some or all of these advantages. In order to ensure that the product is not deficient, or perceived to be deficient, in some way – a role typically played by standards – it may be necessary to apply mitigations. What mitigations should be applied depends on what is lost by not working in accordance with a standard.

Example: If a special flange or fitting is produced, and the bolt circle is reduced to reduce tooling and material costs with a more efficient design, then

1. interchangeability and direct mating with the standard product does not exist. Special adapters could perhaps be developed as a mitigation to allow mating this product with the standard one, but it must be recognized that simple interchangeability does not exist;
2. because the product is not the same trustworthy product that has been sold for a generation or more, credibility is likely reduced. Customers will be wary of the product until it has a sufficiently demonstrated track record. This aspect can be mitigated to some extent by ensuring that the quality control program of the producer (e.g., ISO compliance) is well publicized, by publication of test results, and by a strong sales and marketing effort involving rapid response to any reported problems;
3. the robustness of the design might be questioned, again in light of the track record of the product that is being replaced. A thorough qualification test program and a series of finite element analyses with clear graphics showing reduced stresses, well publicized, might address this problem;
4. lack of known capabilities can be addressed by a suitable series of cut sheets.

One by one, the advantages lost by not having a standard product can be reviewed and addressed. At the end, the reduced costs and the reduced space requirements of the new product might be enough to justify its production anyway. Usually the mitigations involve using a well-defined and well-documented process in place of the standard. Often, they involve one or more company standards in lieu of use of a VCS. Sometimes they involve using as much of the standard as can reasonably be used, taking exceptions only for those aspects for which compliance is impossible, and, even for those aspects, following processes similar to those in the standard itself.

Example: The use of a wind tunnel nozzle in a high temperature wind tunnel. While another standard (the ASME B31.3 Process Piping Code) might be a better fit, the customer wishes to have an ASME BPVC "Code Stamp" so has to ensure quality and have a visible indication of compliance with a recognized standard. The particular tunnel operates at high temperature with high heat flux. In order to control the temperature and ensure that the nozzle does not burn up, a high heat transfer coefficient is required. Only certain materials will do the job, and these particular materials happen not to be listed in the ASME BPVC.

It happens that the copper alloy that was selected as most suitable had in the past been the subject of a code case (the process provided to gain interim approval of designs or products not permitted by the ASME BPVC, pending further consideration and possible inclusion in the code), but the code case was annulled, essentially because of lack of use. The customer has the option of asking for reinstatement of the code case, but this would likely take about a year, delaying critical tests in the wind tunnel.

If the customer cannot either wait or switch to another material, it will not be possible to have an ASME Code stamped nozzle. There are two credible alternatives, neither of which involves a code stamp.

Since the wind tunnel nozzle has a (high) flow through it, one reasonable approach is to design and construct the nozzle in accordance with the ASME B31.3 Process Piping Code, which has provisions for the use of unlisted materials and unlisted components. By following this standard, including the provisions for unlisted materials, a component fully compliant with this standard could be produced (It would not be code stamped because this standard has no provisions for code stamping.).

The other approach would be to follow the ASME BPVC in all respects except for using material and material properties from it. In this situation, because there had previously been a code case for this material, material properties could probably be taken from that code case. Otherwise the material properties could be determined following the same process as is used for listed materials, but since this code does not provide for the use of unlisted materials the component would not be in full compliance with the standard, again resulting in a product that does not have a code stamp.

There does not appear to be any specific regulation that addresses this application, other than the general duty clause of OSHA, so there is no immediate regulatory compliance issue with either approach. Either approach would likely result in a product with adequate structural integrity and that met the needs of the customer. There may be a stronger argument for using the Process Piping Code than for the BPVC for two reasons: first, the wind tunnel nozzle is truly a flow component, so it could be argued that the Process Piping Code is a better fit; second, this standard can be fully complied with. While the component would likely end up with essentially the same design in either case, there is probably a lesser chance of criticism if the standard that can be fully complied with is followed and documented.

A.5   Summary

In some cases, it makes sense to produce products without reference to a standard even if one exists. There are many instances where a standard is not needed or desirable. However, it is important to consider what is being given up by not using an available standard. Any mandatory requirements need to be understood and fully addressed. Caution must be exercised to assure safety and reliability in the product. Issues of interchangeability need to be considered. Customer confidence in the product must be evaluated. Potential increased liability resulting from a "noncompliant" product should be addressed. These aspects are then weighed against any potential benefits.

If reduced cost (sometimes there is no reduction in cost), weight, product lead time, greater flexibility, etc. balanced against any drawbacks of not following a standard still encourage a decision not to use a standard, it may make sense. If this is the decision, then proceed carefully, as if a standard were being applied, documenting decisions and designs so as to get as many of the benefits as possible of using a standard, even without one.

Appendix B

Some SDOs developing Voluntary Consensus Standards

	The Aluminum Association	www.aluminum.org
ABMA	American Bearing Manufacturers Association	www.americanbearings .org
ABMA	American Boiler Manufacturers Association	www.abma.com
ACI	American Concrete Institute	www.concrete.org
ACMA	American Composites Manufacturers Association	www.acmanet.org
AENOR	Spanish Association for Standardization and Certification	www.en.aenor.es
AFNOR	Association Francaise de Normalisation	www.afnor.org/en/
AGMA	American Gear Manufacturers Association	www.agma.org

Primer on Engineering Standards: Expanded Textbook Edition, First Edition.
Owen R. Greulich and Maan H. Jawad.
© 2018, The American Society of Mechanical Engineers (ASME), 2 Park Avenue, New York, NY, 10016, USA (www.asme.org). Published 2018 by John Wiley & Sons Ltd.

AHRI	Air-conditioning, Heating, and Refrigeration Institute	www.ahrinet.org
AIA	Aerospace Industries Association	www.aia-aerospace.org
AIAA	American Institute of Aeronautics and Astronautics	www.aiaa.org
AISC	American Institute of Steel Construction	www.aisc.org
AISI	American Iron and Steel Institute	www.steel.org
AITC	American Institute of Timber Construction	www.aitc-glulam.org
ANS	American Nuclear Society	www.ans.org
API	American Petroleum Institute	www.api.org
ASA	Acoustical Society of America	www.acousticalsociety.org
ASABE	American Society of Agricultural and Biological Engineers	www.asabe.org
ASCE	American Society of Civil Engineers	www.asce.org
ASHRAE	American Society of Heating, Refrigerating and Air-Conditioning Engineers	www.ashrae.org
ASME	American Society of Mechanical Engineers	www.asme.org
ASNT	American Society for Nondestructive Testing	www.asnt.org
ASTM	ASTM International	www.astm.org
ATIS	Alliance for Telecommunications Industry Solutions	www.atis.org
AWC	American Wood Council	www.awc.org
AWEA	American Wind Energy Association	www.awea.org
AWS	American Welding Society	www.aws.org
AWWA	American Water Works Association	www.awwa.org
BIS	Bureau of Indian Standards	www.bis.org.in
BSI	British Standards Institute	www.bsigroup.com
CENELEC	European Committee for Electrotechnical Standardization	www.cenelec.eu
CGA	Compressed Gas Association	www.cganet.com
CGSB	Canadian General Standards Board	http://www.tpsgc-pwgsc.gc.ca/ongc-cgsb/index-eng.html
CISPI	Cast Iron Soil Pipe Institute	www.cispi.org
DIN	Deutsches Institut fur Normung e. V.	www.din.de
EPRI	Electrical Power Research Institute	www.epri.com

EU	European Union	https://ec.europa.eu/ growth/single- market/european- standards_en
GOST	Russian Federal Agency on Technical Regulating and Metrology	http://www.gost.ru/ wps/portal/en
HI	Hydraulic Institute	www.pumps.org
HPVA	Hardwood Plywall and Veneer Association	www.hpva.org
ICC	International Code Council	www.iccsafe.org
IEC	International Electrotechnical Commission	www.iec.ch
IEEE	Institute of Electrical and Electronics Engineers	www.ieee.org
IRAM	Instituto Argentino de Normalizacion y Certificacion	www.Iram.org.ar
ISO	International Organization for Standardization	www.iso.org
JISC	Japanese Industrial Standard Committee	www.jisc.go.jp/eng
KATS	Korean Agency for Technology and Standards	http://www.kats.go.kr/ en/main.do
NACE	National Association of Corrosion Engineers	www.nace.org
NBBI	National Board of Boiler and Pressure Vessel Inspectors	www.nationalboard.org
NEI	Nuclear Energy Institute	www.nei.org
NETA	InterNational Electrical Testing Association	www.netaworld.org
NFPA	National Fire Protection Association	www.nfpa.org
PCA	Portland Cement Association	www.cement.org
SAC	Standardization Administration of the People's Republic of China	http://www.sac.gov.cn/ sacen/
SAE	SAE International (formerly Society of Automotive Engineers)	www.sae.org
SIS	Swedish Standards Institute	http://www.sis.se/en/
TMS	The Masonry Society	https://masonrysociety .org/
UL	Underwriters Laboratories	www.ul.com
UNI	Italian Organization for Standardization	www.uni.com

Appendix C

Some Industrial Organizations That Publish Limited Consensus Standards

Acronym	Organization	Website	Standard(s)
ABMA	American Boiler Manufacturers Association	www.abma.com	Various boiler publications
EJMA	Expansion Joint Manufacturers Association	www.ejma.org	Expansion joint standards
HEI	Heat Exchange Institute	www.heatexchange.org	Heat exchange and vacuum apparatus
TEMA	Tubular Exchanger Manufacturers Association	www.tema.org	Construction of heat exchangers

Primer on Engineering Standards: Expanded Textbook Edition, First Edition.
Owen R. Greulich and Maan H. Jawad.
© 2018, The American Society of Mechanical Engineers (ASME), 2 Park Avenue, New York, NY, 10016, USA (www.asme.org). Published 2018 by John Wiley & Sons Ltd.

Appendix D

Some US Government Jurisdictional Agencies

Acronym	Organization	Website
CPSC	Consumer Product Safety Commission	www.cpsc.gov
DOE	Some US Government Department of Energy	www.doe.gov
DOT	Some US Government Department of Transportation	www.dot.gov
EERE	Office of Energy Efficiency and Renewable Energy	www.energycodes.gov
EPA	Some US Government Environment Protection Agency	www.epa.gov
FAA	Federal Aviation Administration	www.faa.gov
FCC	Federal Communications Commission	www.fcc.gov
FDA	Food and Drug Administration	www.fda.gov
FEMA	Federal Emergency Management Agency	www.fema.gov
FHWA	Federal Highway Administration	https://www.fhwa.dot.gov/
FMCSA	Federal Motor Carrier Safety Administration	www.fmcsa.dot.gov

Primer on Engineering Standards: Expanded Textbook Edition, First Edition.
Owen R. Greulich and Maan H. Jawad.
© 2018, The American Society of Mechanical Engineers (ASME), 2 Park Avenue, New York, NY, 10016, USA (www.asme.org). Published 2018 by John Wiley & Sons Ltd.

Acronym	Organization	Website
FRA	Federal Railroad Administration	www.fra.dot.gov/
FTA	Federal Transit Administration	www.fta.dot.gov/
NHTSA	National Highway Traffic Safety Administration	www.nhtsa.gov
NRC	Nuclear Regulatory Commission	www.nrc.gov
OSHA[a]	Occupational Health and Safety Administration	www.osha.gov
PHMSA	Pipeline and Hazardous Materials Safety Administration	www.phmsa.dot.gov/

[a]A number of states have their own occupational safety and health administrations that have been approved by OSHA. These typically have their own version of the OSHA regulations.

Bibliography

1. AISC *Steel Construction Manual*, 14th Ed. American Institute of Steel Construction.
2. API 579-1/ASME FFS-1 Fitness for Service.
3. ASME *Boiler and Pressure Vessel Code*.
4. ASME B16.34 *Valves – Flanged, Threaded, and Welding End*.
5. ASME *B31 Piping Codes*.
6. Brennan, P. (2013) *Blowback*. The National Board of Boiler and Pressure Vessel Inspectors, Columbus, Ohio.
7. Code of Federal Regulations 29 CFR, *OSHA*.
8. Hunter, R. (2009) *Standards, Conformity Assessment, and Accreditation for Engineers*, CRC Press.
9. ISO/IEC Guide 2. *Standardization and Related Activities – General Vocabulary*.
10. Khan, W. and Raouf, A. (2006) *Standards for Engineering Design and Manufacturing*, CRC Press.
11. OSHA, *Standard Interpretations*, https://www.osha.gov/pls/oshaweb/owasrch.search_form?p_doc_type=INTERPRETATIONS.

Primer on Engineering Standards: Expanded Textbook Edition, First Edition.
Owen R. Greulich and Maan H. Jawad.
© 2018, The American Society of Mechanical Engineers (ASME), 2 Park Avenue, New York, NY, 10016, USA (www.asme.org). Published 2018 by John Wiley & Sons Ltd.

Biography

Owen R. Greulich, M.E., P.E., is a mechanical engineer currently employed as Pressure and Energetic Systems Safety Manager in the Office of Safety and Mission Assurance at NASA Headquarters. His current responsibilities include safety of pressure and vacuum systems while prior work included project management of pressure systems design and implementation, major wind tunnel modifications, fabrication of research hardware, and construction of test stands.

Mr Greulich is responsible for the development and quality of a number of internal standards, including review of standards in a wide range of engineering fields. He is active on the High Pressure Task Group of the American Society of Mechanical Engineers (ASME) Process Piping Code Committee and American Institute of Aeronautics and Astronautics (AIAA) committees on aerospace pressure systems standards. He is Chair of the Fatigue and Fracture subcommittee for aerospace pressure vessels.

Prior to his employment at NASA, Mr Greulich worked in the private sector, designing and managing construction of pressure and vacuum systems and other specialized fabricated and machined hardware. His publications include work on integrity of wiring in aerospace applications, studies regarding rocket propellant safety, and composite structures.

Mr Greulich received his Master of Engineering degree in Mechanical Engineering at the University of California in 1979. He is a registered professional engineer in the State of California.

Primer on Engineering Standards: Expanded Textbook Edition, First Edition.
Owen R. Greulich and Maan H. Jawad.
© 2018, The American Society of Mechanical Engineers (ASME), 2 Park Avenue, New York, NY, 10016, USA (www.asme.org). Published 2018 by John Wiley & Sons Ltd.

Maan H. Jawad, PhD, P.E., is President of Global Engineering & Technology (GE&T). GE&T performs engineering consulting to the pressure vessel industry as well as the power, petrochemical, and nuclear industries. Prior to this, Dr Jawad was on the board of directors and was Director of Engineering of the Nooter Corporation where he was employed from 1968 until he retired in 2002.

Dr Jawad has been active on various technical committees of the ASME Boiler and Pressure Vessel Code since 1972 and has served on numerous committees as a member and chairperson. He was appointed by the Governor of Missouri to the Missouri Board of Boiler and Pressure Vessel Rules and served from 1998 to 2005. He was also a member of the advisory board to the National Board of Boilers and Pressure Vessel Inspectors in Columbus, Ohio, representing The American Boiler Manufacturer's Association from 2000 to 2005.

Dr Jawad obtained his PhD in Structural Engineering from Iowa State University in 1968. He has authored and coauthored five technical books related to pressurized equipment, two of which have been translated into Chinese. He has also authored and coauthored numerous technical research papers in the field of pressure vessels.

Dr Jawad has taught graduate and undergraduate level engineering courses at various universities in topics such as Theory of Plates and Shells, Finite Element Analysis, Advanced Structural Analysis, and Plastic Design. He is a registered professional engineer in the State of Missouri, a Fellow of the ASME, and a member of the American Society of Civil Engineers.

Index

Primer on Engineering Standards: Expanded Textbook Edition, First Edition.
Owen R. Greulich and Maan H. Jawad.
© 2018, The American Society of Mechanical Engineers (ASME), 2 Park Avenue, New York, NY,
10016, USA (www.asme.org). Published 2018 by John Wiley & Sons Ltd.